CYBERSECURITY

CYBERSECURITY

A Self-Teaching Introduction

C. P. GUPTA, PhD
&
K. K. GOYAL, PhD

MERCURY LEARNING AND INFORMATION
Dulles, Virginia
Boston, Massachusetts
New Delhi

Publisher: David Pallai
MERCURY LEARNING AND INFORMATION
22841 Quicksilver Drive
Dulles, VA 20166
info@merclearning.com
www.merclearning.com
1-800-232-0223

C.P. Gupta and K.K. Goyal. *Cybersecurity: A Self-Teaching Introduction.*
ISBN: 978-1-68392-498-2

Library of Congress Control Number: 2020930906

202122321   Printed on acid-free paper in the United States of America.

This book is dedicated to
our families
and
our friends
for making everything possible.

CONTENTS

PREFACE

Considering the risks involved with online freedom and global connectivity, the study of cybersecurity has become an essential component for every computer professional and undergraduate student of computer science. Thus, the prime objective behind publishing this book is to increase the awareness on cybersecurity.

The chapters are organized incrementally by subject. Chapter One is a brief introduction to basic concepts of systems, information systems, threats, and risk. It also gives a general overview of the terms that are going to be used. Chapter Two presents the overview of application layer security per the OSI model of computer networks. It also discusses several secondary data-storage media and their mechanisms, archival, and retrieval systems. Concerns about intrusion detection systems, virtual private networks, electronic transactions, and cryptography are also covered. Chapter Three introduces the implementation of security during each and every phase of the system development life cycle. In this chapter we also discuss the concept of physical security on IT assets. Chapter Four deals with laws related to the cyber world. It emphasizes intellectual property rights, semi-conductor law, and other important acts. The chapter also provides the do's and don'ts in IT.

In the end, our hope is that you will be in a better position to make informed decisions about anything of consequence that might be affected by the growing potential for cyber-attacks. This book should appeal to a wide audience as an effective self-study guide. Nothing is possible without expert guidance and improvement, so please send us feedback.

Dr. C.P. Gupta
Dr. K.K. Goyal
February 2020

CHAPTER 1

INTRODUCTION TO INFORMATION SYSTEMS

1.1 INTRODUCTION TO INFORMATION SYSTEMS

1.1.1 What Is Information?

Data versus Information

Data is a stream of raw facts representing events such as business transactions. Once this data is processed, it can more easily be turned into information.

Data that has been put into a meaningful and useful context, usually called information, helps to make a decision. *Applied formation* is a cluster of facts that is meaningful and useful to human beings and help in the process of making decisions.

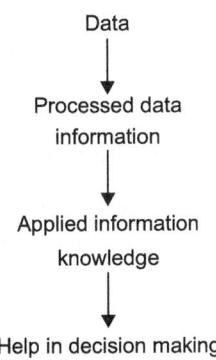

Data
↓
Processed data
information
↓
Applied information
knowledge
↓
Help in decision making

"Information Is Power"

The goal is to put the right information, in the right hands, at right time, in the right format.

Good information is useful to the decision maker, and it must meet certain criteria. Some of the *characteristics of good information* are confidentiality, integrity, availability, possession, utility, and accuracy.

Information Classification

Information can be classified in terms of usefulness, age, and value according to the requirements for protection. Information is always updated and if new information is found, then *usefulness* defines the level of importance. The importance of information also depends on time and how long the information is valid if time (i.e., age is passed then importance of information is zero). The *value* of information is very important for private areas.

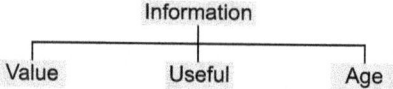

Users According to Information Classification

According to information classification, information is an asset. Users of information have the responsibility to protect the data. Those *users* and *their roles and responsibilities* are defined as follows:

An *owner* is an individual who is responsible for establishing the policy and makes final decisions. He keeps the information secret to protect from corresponding levels of classification. The owner also manages the custodian and user. A *custodian* is an individual responsible for maintaining the information by taking backups. He keeps the middle level of information that is sensitive and secret. There are end *users* or operators who collect raw data for information and follow operating procedures.

1.1.2 What Is a System?

A system is a set of activities that have been arranged systematically to achieve a certain objective. A system is a set of related components that can process an input to produce a certain output. Every system requires a form of data input, like an ATM machine accepts data when we enter PIN number and a washing machine accepts data when we select start buttons. They process inputs and produce respective outputs.

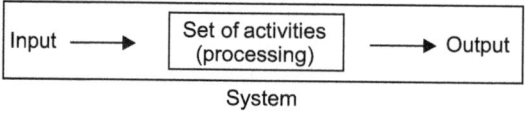

1.1.3 What Is an Information System (IS)?

An information system (IS) is a combination of information technology, people, and activities that support the operations of management and decision making. In other words, an information system is not simply about computer—it's about how businesses can make the best use of computer technology to provide the information needed to achieve the goals.

An information system is used to provide the interface between people, processes, data and technology, which can support business operations to increase productivity and help managers in making decisions.

IS is a combination of information technology plus a set of activities to serve as a support for decision making.

An Information System Contains Five Major Components

1. **Hardware:** Hardware is a part of a system or information system that can be touched and seen (i.e., it is physically available and helps in obtaining desired output). Information systems' hardware refer to all type and are used for inputting, processing, managing, solving, and distributing the information being used in an organization, such as physical computers, networks, communication equipment, scanners, digital drives, and so on.

$$I/P \Leftrightarrow Processing \Leftrightarrow O/P$$
$$\Updownarrow$$
$$Storage$$

2. **Software:** Software is the part of computer system we cannot see and feel. It does not have physical existence—we cannot touch it—but it is logically available: Its presence can be felt in performance of certain tasks.

Software can be broadly classified into two categories:

- System software
- Application software

System software is a special type of software that controls the device driver that can communicate with specific hardware whenever required. Application software contains the programs that can help users and enable companies to perform business functions. Users can increase productivity with the presence of application software such as Microsoft Word, Tally, Paint, Adobe Photoshop, and so on.

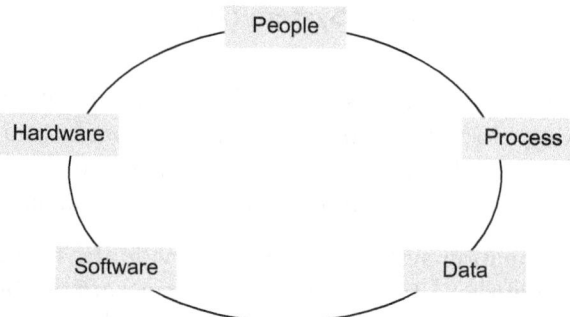

3. **Data:** Data refers to raw facts and figures on any subject or entities such as student names, courses, and marks. Raw data can be processed to become more useful information.

   Information is an organized, meaningful, and useful interpretation of data such as a company's performance or a student's academic performance. Information systems change data into information, which gives a certain meaning to its users.

4. **Process:** Process or procedure explains the activities carried out by users, managers, and staff. Process is important for supporting certain business models in written documents or as reference material online. It is a guide consisting of orderly steps to be followed and implemented in order to get a certain decision on a certain matter.

5. **People (humans):** The main objective of an information system is to provide valuable information to managers, users, and others, whether inside or outside of the company. This includes not only users, but those who operate and service the computer, those who maintain the data, and those who support the network of computers. Users can be broken up into three categories:

   – End users
   – Internal users
   – External users

History of Information System (IS)/Information Is Power

Year	Main Activities	Skill Required
1970s	– **Mainframe computers** were used – Systems were tied to a few business functions: **payroll, inventory, billing**	**Programming in COBOL**

(continued)

(continued)

Year	Main Activities	Skill Required
	– Main focus was to **automate existing process** – **PC and LAN** were installed – Departments set up their **own computer**	
1980s	– **End-user computed** with word processor – **Main focus was to automate** existing process	**PC support and basic networking**
1990s	– **Commercialization** – **WWW** (World Wide Web) becomes corporate standard – **System and data integration** – **No-more standalone system** – WWW extended by **Internet**	**N/W support system integration DBA**
2000s	– **Data sharing** – Main focus on efficiency of system (e.g., **speed, manufacturing, distribution**)	**N/W support DBA system integration**

Dimensions of Information Systems

An information system represents the combination of management, organization, and technology. Using information systems effectively requires an understanding of the organization, management, and information technology dimensions of the system. These dimensions can be explained as follows:

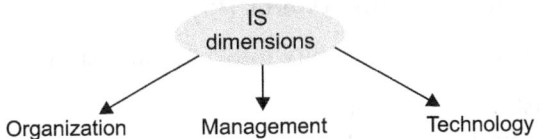

Organization dimension: An information system plays a vital role for managing an organization. This dimension involves the hierarchy of authority, procedures, policies, and operations. Today, with the development of Internet, technology and telecommunications information systems are involved in every phase of organizations such as production, manufacturing, human resources, finance, sales, and marketing.

Management dimension: Organizations set goals, strategies, and plans, and they are handled by different individuals according to their managerial level (top-level management, middle-level management, and low-level management).

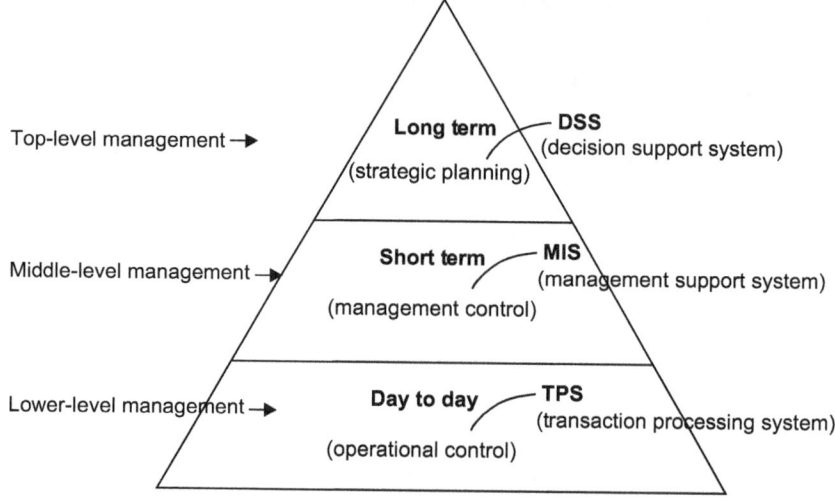

Business organization involves three levels of authority

Information systems provide different tools and information needed by managers to allocate, coordinate, and monitor their work, make decisions, and create new products and reports.

Technology dimension: Technology acts as a tool for information systems; it provides IT infrastructure a platform. It consists of computer hardware and software, data management technology, and networking and telecommunications technology (networks, Internet, intranets, and extranets). Managers use technology to carry out their work and make decisions.

Characteristics of Information Systems (IS)

Characteristics of good quality information systems (IS) are as follows:

- Consistent information
- No redundancy in information
- Updated information
- Complete information
- Appropriate information
- Valid information
- Reliable information

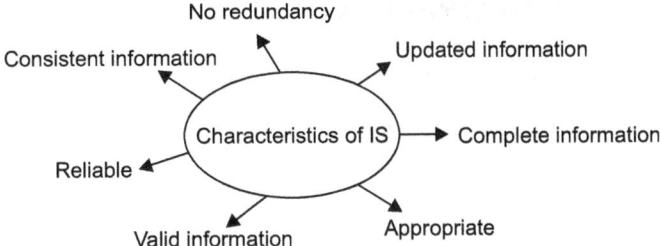

- **Consistent information:** Information should be correct and unambiguous. It should not contain any logical contradiction. Management must be able to rely on the results generated by information systems. Only then can they make good decisions; if information systems produce unreliable information, it could affect the reputation and business of the organization.

- **No redundancy in information:** Information systems ensure no redundancy of data. Redundancy refers to unnecessary, surplus, duplicate information. It can be controlled by removing duplicate data from databases or by taking backup. Redundancy in information systems is simply the result of poor planning.

- **Updated information:** Information is an asset and has its own importance and value over a period of time. The latest updated information plays a crucial role in decision making and planning. Information should be timely updated and communicated so that appropriate action can be taken. The latest events should be updated in information systems as early as possible.

- **Complete information:** Accuracy of information is just not enough. It should also be complete, which means facts and figures should not be missing or concealed. Telling the truth but not wholly is of no use.

- **Appropriate information:** Information should be understandable to the users. It should be communicated according to the structure, format, and details that are relevant to the user. Managers need detailed, summarized reports for appropriate understanding the status of work at a glance.

- **Valid information:** Completeness of information is just not enough. It should be accurate based of facts and figures generated by the amount of research done. It depends on the qualification and experience of the person communicating the information.

- **Reliable information:** Information should be fair and free from bias. It should always yield the same result. To increase the reliability of information, it should be taken directly in writing rather than taken indirectly so that the source and reliability of information can be trace out.

1.2   TYPES OF INFORMATION SYSTEMS

Information systems can be broadly classified in to two categories:

- Management support system
- Operational support system

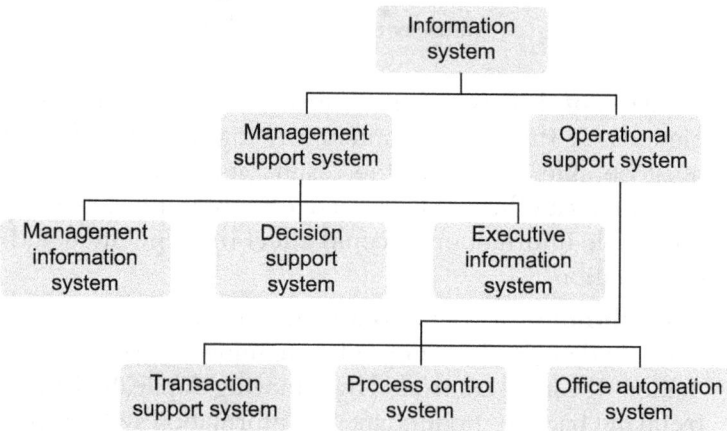

1. **Transaction processing systems (TPS):** TPS were introduced during 1950–1960. TPS is a computerized system that performs and records daily routine transactions necessary to conduct business. TPS process data from business transactions, for example, payroll systems and production instruction.

   There are five stages of TPS:

   (i.)   Data entry
   (ii.) Processing
   (iii.) Database maintenance
   (iii.) Inquiry processing
   (iv.) Document and report generation

2. **Management information systems (MIS):** MIS were introduced during 1960–1970. Management information systems provide information for managing an organization. They extract summarized data from TPS. An MIS allows managers to monitor and direct the organization's processes and activities. It provides accurate feedback and pre-specified reports on a scheduled basis. The output is often of the kind that you need routinely each term (quarterly, monthly, yearly) to evaluate how to proceed.

3. **Decision support systems (DSS):** DSS were introduced during 1970–1980. A decision support system is an interactive information system that provides information, models, and data manipulation tools to help in making decisions in semi-structured and unstructured situations. It supports analytical work, simulation and optimization. Simulation models calculate the simulated outcome of tentative decisions and assumptions. Optimization models determine optimal decisions based on criteria supplied by the user, mathematical search techniques, and constraints. For example, *online analytical processing (OLAP)* provides the use of data analysis tools to explore large databases of transactional data. *Data mining* provides the use of analysis tools to find patterns in large transactional databases.

4. **Executive information systems (EIS):** EIS was introduced during 1980–1990. It is a highly interactive information system that provides flexible access to information for monitoring results and general business conditions and uses both internal and competitive information. It is also known as an *executive support system (ESS)*.

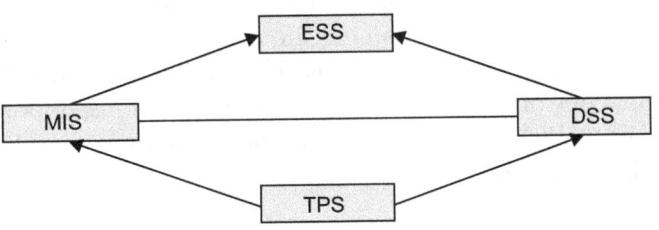

5. **Expert systems (ES):** ES were introduced during 1990–2000. Expert systems support professionals faced with complex situations requiring expert knowledge in a well-defined area. They represent human expertise, also called *knowledge-based Systems* that typically use if-then rules. They are used as interactive advisors or as automated tools.

6. **Office automation systems (OAS):** Office automation systems help people in performing personal record keeping, writing, and calculations efficiently. Main types of tools include:
   - spread sheet programs
   - text- and image-processing systems, and
   - personal databases and note-taking systems.

1.3   DEVELOPMENT OF INFORMATION SYSTEMS

Information system development involves various activities performed together. The stages involved in the IS development life cycle are as follows:

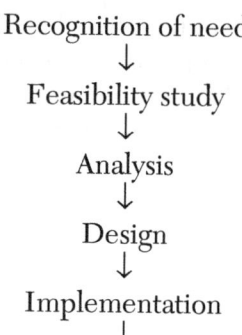

Recognition of need
↓
Feasibility study
↓
Analysis
↓
Design
↓
Implementation
↓
Post implementation and maintenance

1. **Recognition of need:** Recognition of need is the first stage of the information system development life cycle. This gives a clear picture of what the existing system actually is. The preliminary investigation must define the scope of project and the perceived problems, opportunities, and directives that triggered the project. The preliminary investigation includes the following tasks:

    (i.)   Lists problems, opportunities, and directives
    (ii.)   Negotiates project worth
    (iii.) Accesses project worth
    (iv.)   Plans the project
    (v.)   Presents the project and plan

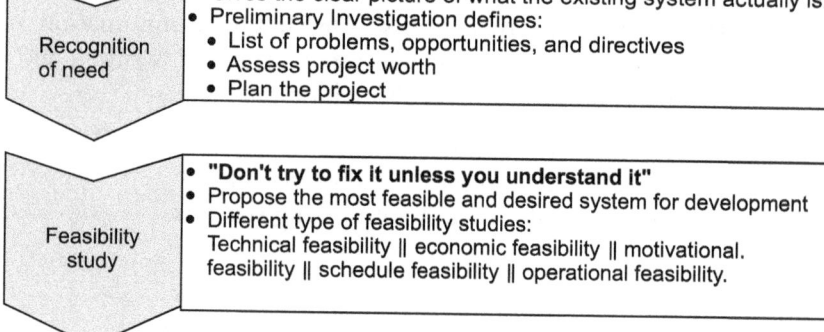

Recognition of need
- Gives the clear picture of what the existing system actually is.
- Preliminary Investigation defines:
  - List of problems, opportunities, and directives
  - Assess project worth
  - Plan the project

Feasibility study
- **"Don't try to fix it unless you understand it"**
- Propose the most feasible and desired system for development
- Different type of feasibility studies:
  Technical feasibility || economic feasibility || motivational. feasibility || schedule feasibility || operational feasibility.

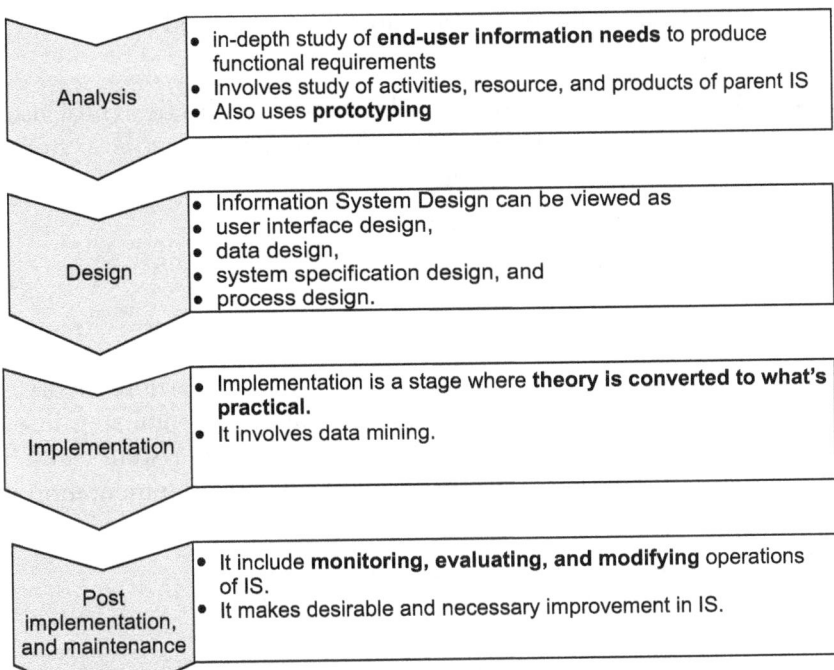

Analysis
- in-depth study of **end-user information needs** to produce functional requirements
- Involves study of activities, resource, and products of parent IS
- Also uses **prototyping**

Design
- Information System Design can be viewed as
- user interface design,
- data design,
- system specification design, and
- process design.

Implementation
- Implementation is a stage where **theory is converted to what's practical.**
- It involves data mining.

Post implementation, and maintenance
- It include **monitoring, evaluating, and modifying** operations of IS.
- It makes desirable and necessary improvement in IS.

2. **Feasibility study:** The statement, *"Do not try to fix it unless you understand it"* applies and completely describes the feasibility study stage of system analysis. The goal of feasibility studies is to evaluate alternative systems and to propose the most feasible and desirable system for development. It consists of a statement of the problem, a summary of the findings, details of the findings, and recommendations and conclusions. There are five types of feasibility studies:

- Technical feasibility
- Economical feasibility
- Motivational feasibility
- Schedule feasibility
- Operational feasibility

3. **Analysis:** Analysis is not a preliminary study. It is an in-depth study of end-user information needs that produces functional requirements that are used as the basis for the design of a new information system. IS analysis involves detailed study of the following:

- The information needs of the organization and end user
- The activities, resources, and product of any present information system

- The information system capabilities required to meet the information needs

4. **Design:** System design can be viewed as the design of user interface, data, process, and system specifications. Information system design can be viewed as the following:

   - User interface design
   - Data design
   - System specification design
   - Process design

5. **Implementation:** Implementation is the stage where theory is converted into what's practical. The implementation is a vital step in ensuring the success of new systems. Even a well-designed system can fail if it is not properly implemented. Information system implementation involves the following:

   - Acquisition of H/W, S/W, and services
   - Software development and modification
   - End-user training
   - System documentation

6. **Post implementation and maintenance:** Once a system is fully implemented and being operated by an end user, the maintenance function begins. System maintenance is the monitoring, evaluating, and modifying of operational information systems to make desirable or necessary improvements.

1.3.1 Prototyping

Prototyping is a part of the analysis phase of the IS development life cycle. It is the process of building a model system (i.e., simulation). Information system prototypes are employed to help system designers build an information system that is intuitive and easy to manipulate by end users. Prototyping is an iterative process.

Advantages of Prototyping

- Prototype reduces the development time.
- It reduces the development cost.
- It requires user involvement.
- Its results provide higher user satisfaction.

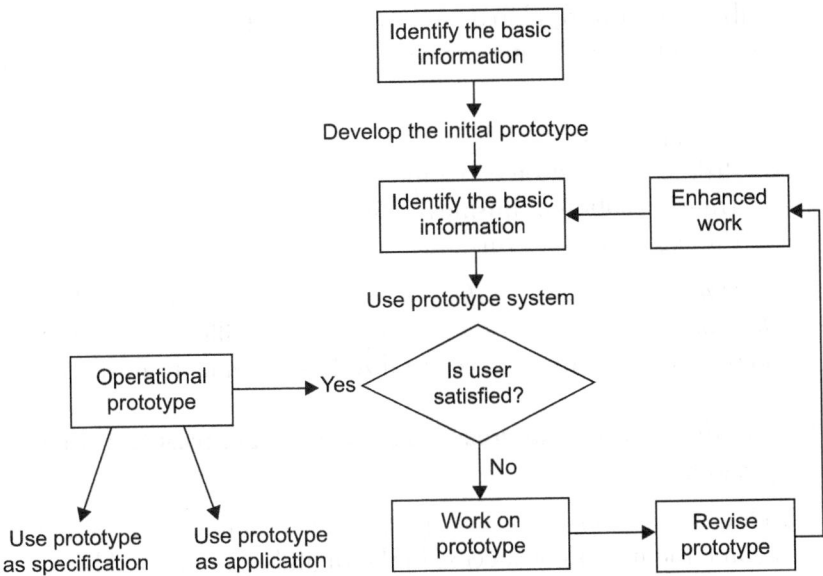

- Prototyping is an active, not passive model that end users can see, touch, and feel.
- Prototyping can increase creativity because it allows for quicker user feedback, which can lead to better solutions.

Disadvantages of Prototyping

- Prototyping often leads to premature commitment to a design.
- Prototyping can reduce creativity in designs.
- Not suitable for larger applications.
- Structure of the system can be damaged since many changes can be made.
- The developer can misunderstand users' objectives.

1.4   CHANGING THE NATURE OF INFORMATION SYSTEMS

Information system changes should be timely and done periodically. It is because of globalization, transformation of information in digital world. Each information system is defined and characterized by their purpose and goal. Today, systems are not residing in a single location but are distributed geographically and based on component. Today information systems are flexible, structured, global, and integrated. They are also very fast and accurate. Business environments have been changed or altered by four powerful changes:

– Globalization
– Risk in information economy
– Transformation of business enterprise
– Emerging digital firms

Main frame–based information system changes to client server–based information system. This system improves usability, flexibility, interoperability, and scalability compared to centralized main frame time-sharing computing.

1.4.1 Globalization of Business and Need of Distributed Information Systems

Today, businesses have no geographical boundaries as globalization has become the mantra of success in the digital economy, led by rise of electronic business (e-business). "Extended enterprises" are completely new ways of doing business, e-commerce, or e-business. Information systems (IS) are handling business in all forms, not just the text-based data of 1970s that typically comes in flat files but also the rich text, image, graphics, and voice. There are digital markets where an IS links buyers and sellers to exchange information products, services, and payments worldwide.

Role of Internet and Web Servers in Globalization of IS

The Web is huge collection of multimedia information located on Web servers attached to Internet. Web servers have proved to give a strong return on investment (ROI) and make computer-based IS more reliable. They also bring productivity, flexibility, and low maintenance costs in the development of IS by integrating components from various third-party venders. Linking information is very easy; it can be shared all around the world.

1.4.2 Needs in Distributed Information Systems (DIS)

The distributed development of information systems (also named global software development) increased with the globalization of companies and their businesses with new information and communication technologies. In distributed information systems, the information is available on several computers (i.e., information is distributed over multiple sites, which are connected and share information with a communication network). In distributed information systems, many computers connect to each other and share their information with each other. It increases enterprise productivity, reducing IS development cost and extending enterprise strategy to the global market. Certain common characteristics needed in distributed information systems are the following:

1. **Transparency:** A distributed system needs to hide the fact that its processes and resources are physically distributed across multiple computers. Transparency is of various forms, as follows:

Transparency	Description
Access	Hide differences in data representation and how a resource is accessed
Location	Hide where a resource is located
Migration	Hide that a resource may move to another location
Relocation	Hide that a resource may be moved to another location while in use
Replication	Hide that a resource is replicated
Concurrency	Hide that a resource may be shared by several competitive users
Failure	Hide the failure and recovery of a resource

2. **Easier expansion:** Distributed information systems (DIS) allow many users to access common data. DIS share the workload over available machines and enhance sharing and processing of information. DIS are always open to add new resources and sharing services. They remain effective when there is significant sharing of resource and users.

3. **Reflects original system and local autonomy:** This system enables information to be accessed without knowing their location. Data can be placed at different sites users normally use. In this way, users have local control of the data, and they can establish and enforce local policies regarding the use of this data.

4. **Protection of valuable data:** Distributed information systems are responsible for protecting valuable data even if any fault or failure occurs. It can be achieved by the recovery and redundancy of information or valuable data.

5. **Economic:** It creates network of small computers with the power of single large computer at less cost. A collection of computers offers better

price/performance than a single mainframe computer. It's the cost-effective way to increase computing power.

6. **Modularity:** Systems can adapt new features and can be deleted easily, which does not affect the other system. Computing power can be added in small increments. This leads to modular expandability.

7. **Reliable transaction:** Due to replication of data and secure channels with reliable networks, we can manage large and complex transactions very easily. If any machine crashes the system can still survive. It ensures availability and improves reliability. As data may be replicated so that it exists at more than one site, it makes data available, even the failure of a node or a communication link.

Characteristics of Distributed Information Systems (DIS)

- **Communication:** The computers in a distributed system communicate with one another through various communication media, such as high-speed networks or telephone lines. They do not share main memory or disks.

- **Distributed data:** Data sets can be split into fragments and can be distributed across different nodes within a network. Individual data fragments can be replicated and allocated across different nodes.

- **Performance:** Each site is capable of independently processing user requests that require access to local data or file systems and is also capable of performing and processing on remote machines in the network.

- **Distributed operations:** Distributed systems operate in parallel; different algorithms ensure the correctness, especially operation during failures of part of the system, and recovery from failures.

Disadvantages of Distributed Information Systems (DIS)

- **Complex:** The added complexity required to ensure proper coordination among the sites is the major disadvantage. This increased complexity takes various forms like cost, bugs, and processing overheads.

- **Security:** There may be security problems due to the sharing nature of distributed systems.

- **Difficulty to maintain integrity:** It is very difficult to maintain the integrity because some messages can be lost in the network system.

- **Additional hardware and software required:** Additional hardware and software are required to manage bandwidth and overloading problems and to increase performance and network operations.

1.5   INTRODUCTION TO INFORMATION SECURITY

A modern society is heavily dependent on information and communication technologies. We trust computers, but they are often not safe or secure and therefore have many risks. *"The major goal of information security is to define ways to reduce these risks."* Said an other way, the objective is to protect against those who do harm intentionally or un-intentionally. Security is the quality or state of being secure to be free from danger. Fear of anything that can go wrong is *threat* and the possibility of anything going wrong is called *risk*. Action for doing anything wrong is called *attack*. We need to secure form threats. Multilevels of security were implemented by "onion skin" or "defense in depth" approach:

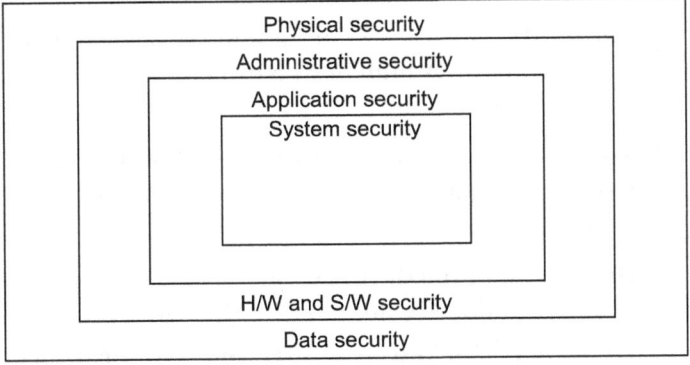

Security Layers

Information security protects information *assets*. Assets may be physical or logical assets. Physical assets are tangible like hardware, people, and computer systems and logical assets are intangible like data, information, and websites. Assets can be further classified into information, software, hardware, people, and systems.

There are three things that may help to protect information systems:

- **Education:** Education of what could happen if lost or in the wrong hands

- **Awareness:** Awareness of what information you have. How important it is? How secure it is?

- **Technology:** Technological precautions to secure information systems

1.5.1 Key Concepts of Information Security

Information security has some key concepts to secure or protect valuable assets:

- **Access:** Authorized users have legal access to the system, whereas hackers have illegal access to the system. Access control regulates this.

- **Asset:** Assets can be physical or tangible like the organizational resource that is being protected. Assets may be logical such as websites or may be any information/data that needs to be protected to identify the value.

- **Attack:** An intentional or unintentional act that can cause damage or otherwise compromise information or the systems. Attacks can be active or passive, intentional, or unintentional and direct or indirect.

- **Vulnerability:** A weakness or fault in the system that can be exposed to cause loss or harm. A human who exploits vulnerability is responsible for the attack on the system.

- **Threat:** Set of circumstances that causes loss, harm, or danger to an asset. A weakness or fault in a system or protection mechanism that opens it to attack or damage.

- **Control or safeguard:** An action, device, procedure, or technique that eliminates or reduces vulnerability. Also called a countermeasure.

  || **Threats are blocked by controlling vulnerability.** ||

1.6   NEED FOR INFORMATION SECURITY

The value of information comes from the characteristics it possesses. When a characteristic of information changes, the value of that information either increases or decreases. Some characteristics affect information's value to users more than others do, called essential characteristics of the information.

The need for information security thus arises to maintain confidentiality, integrity, and availability—three essential characteristics of information. This is defined as the CIA triangle, CIA triad, or security triad of information security.

Confidentiality: Confidentiality is the "prevention of disclosure of information from unauthorized user." Information has confidentiality when it is protected from disclosure or exposure to unauthorized individuals or systems.

Confidentiality ensures that only those with the rights and privileges to access the information are able to do so. When unauthorized individuals or systems can view information, confidentiality is breached. To protect the confidentiality of information, we can use a number of measures, including the following:

- Information classification
- Secure document storage
- Application of general security policies
- Education of end users

Integrity: Integrity is the "prevention of modification of information from unauthorized user." Information has integrity when it is whole, complete, or uncorrupted. The integrity of information is threatened when the information is exposed to corruption, damage, or destruction, or when its authentic state is disrupted. Corruption can occur while information is being stored or transmitted. Many computer viruses and worms are designed with the explicit purpose of corrupting data. For this reason, a key method for detecting a virus or worm is to look for changes in file integrity as shown by the size of file.

Availability: Availability is the "right information accessed by right person at the right time." Availability enables authorized users, people, or computer systems to access information without interference or obstruction and to retrieve it in the required format.

CIA triangle:

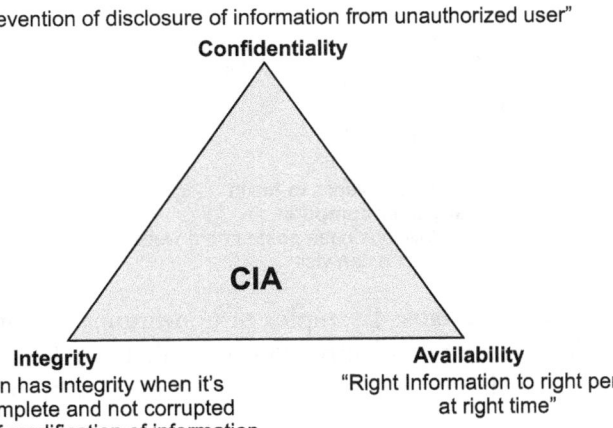

1.7 THREATS TO INFORMATION SYSTEMS

"A *threat* is a possible event or danger that can harm an information system"; it's fear that something is wrong, and the probability of something being wrong is *risk*. The action of doing anything wrong is an *attack*.

"*A threat can be controlled or blocked by controlling vulnerability.*" We can consider potential harm to assets in two ways: We can look at what bad things can happen to assets or we can look at who or what can cause or allow those bad things to happen. These two perspectives enable us to determine how to protect our assets. Threats can be defined as categories of objects, people, or entities that present danger to an asset. Threats are always present and can be purposeful or undirected. *Vulnerability* can be defined as a weakness or fault in a system.

One way to analyze harm is to consider the cause or source. We notify the potential cause of harm or a threat.

Source of Threats

Natural disasters
– Fire/floods
– Electric power break down
– Failure of communication
– Failure of disk drive

• Active (I/Un)/passive(I)
• Direct (I/Un)/indirect(I)
• Intentional/unintentional

Threats

Natural causes   Human causes

(Non-malicious)
Unintentional (human error)
– Spilling of drink on laptop
– Deleting text
– Sending email to wrong person
– Typing 12 instead of 21
– Clicking yes instead of no

Intentional
– Attacker intends harm

Benign intent   Malicious intent

Random   Directed

Attacker wants to harm any user/computer
– Malicious code posted on a website anyone can visit

Attacker harm to specific user/computer
– Account transaction

Natural causes: Examples of non-human threats include natural disasters like fires, floods, electric power break down, failure of equipment or component like disk drives, and failure of communication channel.

Human causes: Human threats can either be cause of *benign intent or malicious intent*.

Benign threats are also called non-malicious threats caused by humans unintentionally. They include kinds of harm like someone accidentally spilling a soft drink on a laptop, unintentionally deleting a file, inadvertently sending an email message to the wrong person, typing 12 instead of 21, or clicking yes instead of no.

Malicious threats are caused by humans intentionally. They are attacks and can be categorized as intentional, unintentional, active, passive, direct, or indirect. Malicious attacks can be *random or directed*.

In a *random attack* the attacker wants to harm any computer or user; such an attack is analogous to accosting a pedestrian who walks down the street. An example of a random attack is malicious code posted on a website that anyone can visit.

In a *directed attack*, the attacker intends harm to specific computers, perhaps at one organization or a specific individual account transaction.

1.7.1 Classification of Threats on the Basis of Damages

On the basis of assessing damage, threats can be classified in term of assets, actors, motives, access, and outcome.

Assets: Things that have the value to the organization. The organizational resources that are being protected are called assets. An asset can be logical or physical. Logical assets are intangible like data, information, websites, and software. Physical assets are tangible like hardware, people, and computer systems.

Information is preserved in the form of papers or electronic documents and need to be protected from damage. Softwares are licensed; most software give access to authorized user only, so there is a need to protect from unauthorized access. In information technology, all the information is stored and travels in different types of hardware like storage devices and communication devices, so the replacement cost from damage needs to be considered.

1.7.2 Life Cycle of Threat

Attackers use system vulnerability to generate different types of damages. These attackers are called agents for threat. Threat agents give rise to threats by exposing vulnerability or weakness in the system, which leads to risk that damages assets in the information system. The threat can be encountered or exposed by applying various safeguard strategies.

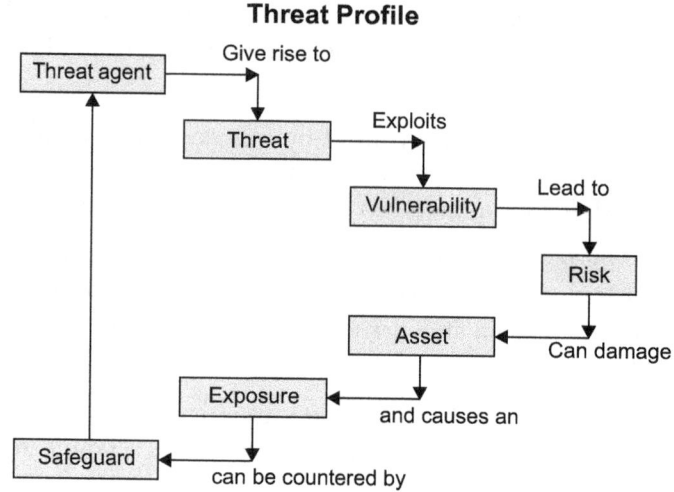

Threat Profile

1.7.3 Threats Related to Information Systems

S.No.	Name	Description
1.	Impersonation	Enjoy the privileges of a legitimate user by gaining access to system
2.	Trojan horse	Program that contains unexpected functionality
3.	Logic bomb	Triggers action when condition is encountered; code can be enabled in any program.
4.	Computer virus	Approaches itself toward a program and propagates its copies to other programs
5.	Denial of services (DoS)	Disturbs the services provided by the server
6.	Dial diddling	Changing data before or during input, often to change the content of database
7.	Salami attack	Diverting a small amount of money from a large number of accounts maintained by the system
8.	Spoofing	Configuring a computer to masquerade as another system over the network in order to get unauthorized access to the resources of the system being attacked.

(continued)

S.No.	Name	Description
9.	Theft of mobile device	—
10.	Zombie	Program that activated on an infected microprocessor that is activated to launch attacks on an other microcontroller
11.	Worm	Program that propagates copies of itself to another system or network
12.	Write tapping	—
13.	Data leakage	—
14.	Super zapping	Using a system program that can pair regular systems' controls to perform unauthorized access
15.	Sequencing	Searches the Context from the memory available to the targeted computer.

1.8   INFORMATION ASSURANCE

Information is about ensuring that authorized users have access to authorized information at authorized times. Information assurance ensures that information is 100% secure when it is being processed, because it follow all principles of information security. These principles protect and define the information system by assuring availability, integrity, confidentiality and non-repudiation of information.

Information assurance defines and applies the policies and standards, methodologies, services, and mechanisms to maintain integrity with respect to people, process, technology, information, and supporting infrastructure. Information assurance confirms all characteristics of the secure information system.

Major Aspects of Information Assurance

- Business continuity
- Information security
- Governance
- Data recovery
- Privacy

Core Principles of Information Assurance (IA)

- **Confidentiality:** Confidentiality is the "prevention of disclosure of information from unauthorized users." Information has confidentiality when it is protected from disclosure or exposure to unauthorized individuals or systems. Confidentiality ensures that only those with the rights and privileges to access the information are able to do so. When unauthorized individuals or systems can view information, confidentiality is breached. To protect the confidentiality of information, we can use a number of measures including the following:

  - Information classification
  - Secure document storage
  - Application of general security policies
  - Education of end users

- **Integrity:** Integrity is the "prevention of modification of information from unauthorized users." Information has integrity when it is whole, complete, or uncorrupted. The integrity of information is threatened when the information is exposed to corruption, damage, destruction, or the disruption of its authentic state. Corruption can occur while information is being stored or transmitted. Many computer viruses and worms are designed with the explicit purpose of corrupting data. For this reason, a key method for detecting a virus or worm is to look for changes in file integrity as shown by the size of file.

- **Availability:** Availability is the "right information accessed by right person at the right time." Availability enables authorized users, people, or computer systems to access information without interference or obstruction and to retrieve it in the required format.

- **Authenticity:** Authenticity of information is the quality or state of being genuine and original. Information is authentic when it is in same state in which it was carried, stored, placed, and transformed.

- **Possession:** The possession of information is the quality or state of ownership and control. Information is said to be in one's possession if one obtains it independent of format and other characteristics. A breach of confidentiality always results in breach of possession, not vice versa.

- **Utility:** The utility of information is the quality or state of having a value for some purpose. Information has a value when it can serve a purpose. If information is available, but it is not in a meaningful format to the end user, it is not useful.

- **Privacy:** Information privacy is the privacy of personal information and usually relates to personal data stored on computer systems. Information privacy is also known as data privacy. The need to maintain information privacy is applicable to collected personal information, such as medical records, financial data, criminal records, political records, business-related information, or website data. Information systems require protecting and ensuring privacy of information.

- **Non-repudiation:** Non-repudiation is the assurance that someone cannot deny something.

Information assurance process:

1. Classification of information assets
2. Risk assessment, analyzing vulnerability and threats
3. Risk analysis
4. Risk management (i.e., treatment against the risk)
5. Test and review
6. Repeat whenever required

1.9 CYBERSECURITY AND SECURITY RISK ANALYSIS

Cybersecurity refers to preventative methods used to protect information from being stolen, compromised, and attacked. It requires an understanding of potential information threats, such as viruses and other malicious code.

Cybersecurity strategies include identity management, risk management, and incident management. They are measures taken to protect computer or computer systems (as on Internet) against unauthorized access or risk refer as cybersecurity.

The Internet allows users to gather, store, process, and transfer vast amounts of data, including proprietary and sensitive business, transactional, and personal data. At the same time that businesses and consumers rely more and more on such capabilities, cybersecurity threats continue to plague the Internet economy.

Cybersecurity threats evolve as rapidly as the Internet expands, and the associated risks are becoming increasingly global. Staying protected against cybersecurity threats requires all users, even the most sophisticated ones, to be aware of the threats and to improve their security practices on an ongoing basis. Creating incentives to motivate all parties in the Internet economy to make appropriate security investments requires technical and public policy measures that are carefully balanced to heighten cybersecurity without creating barriers to innovation, economic growth, and the free flow of information.

As leaders of our organization, we are responsible for protecting the information in our care. Cybersecurity is a business function and technology tool that can be used to more securely protect information assets.

Top Ten Cybersecurity Action Items

1. Designate a principal individual responsible for cybersecurity.
2. Know how to recognize there might be a problem.
3. Understand how to deal with the problems.
4. Physically protect equipment.
5. Protect essential software and hardware.
6. Control access.
7. Protect information.
8. Implement training and awareness programs.
9. Develop Internet and acceptable use policies.
10. Take steps to secure disposal of storage media and equipment.

1.9.1 Security Risk Analysis

A particular threat can exploit a particular vulnerability. Understanding risks for a system and deciding how to control them is necessary.

Risk Management Cycle

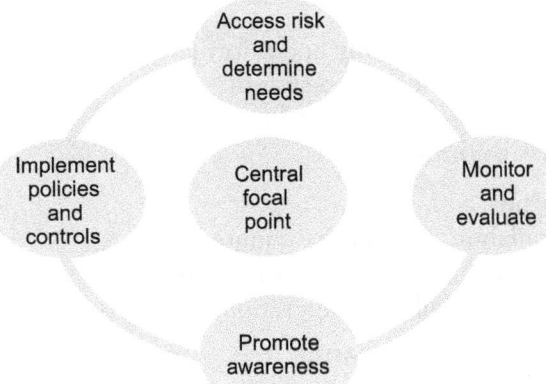

What is risk analysis? The process of identifying, assessing, and reducing risks to an acceptable level includes defining and controling threats and vulnerabilities and implementing risk-reduction measures. It's an analytic discipline with three parts:

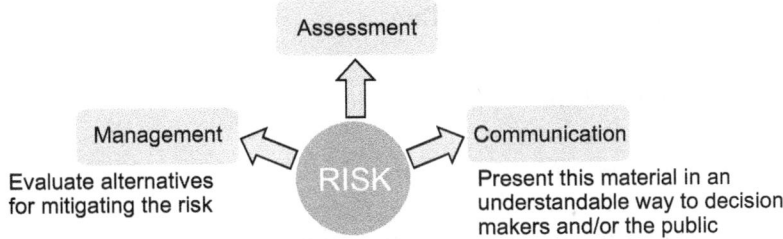

- **Risk assessment:** Determine what the risks are.
- **Risk management:** Evaluate alternatives for mitigating the risk.
- **Risk communication:** Present this material in an understandable way to decision makers and to the public.

Benefits of Risk Analysis

- Assurance that greatest risks have been identified and addressed
- Increased understanding of risks
- Mechanism for reaching consensus
- Support for needed controls
- Means for communicating results

Basic Risk Analysis Structure

- **Evaluate**
  - Value of computing and information assets
  - Vulnerabilities of the system
  - Threats from inside and outside
  - Risk priorities

- **Examine**
  - Availability of security countermeasures
  - Effectiveness of countermeasures
  - Costs (installation, operation, etc.) of countermeasures

- **Implement and monitor**
- **Identify assets**

Types of Risk Analysis

1. **Quantitative:** Quantitative risk is measurement of data that can be put into numbers. The goal of quantitative measurement is to run statistical analysis, so data has to be in numerical form.

   Quantitative Risk Outline

   - Project level
   - Probabilistic estimates of time and cost
   - Time consuming
   - May require specialized tools
   - Identify and value assets
   - Determine vulnerabilities and impact
   - Estimate likelihood of exploitation
   - Compute annual loss exposure (ALE)
   - Survey applicable controls and their costs
   - Project annual savings from controls

2. **Qualitative:** Qualitative risk focuses on collecting information that is not numerical. You can remember this by thinking of the word "quality." Quality is not something that you measure with numbers. An organization's risks and threats are based on judgment, institution, and experience. Rank seriousness of the threats for the sensitivity of the assets.

   Qualitative Risk Analysis

   - Risk level
   - Subjective evaluation of probability and impact

- Quick and easy to perform
- No special software or tools required
- Hard to make meaningful valuations and meaningful probabilities
- Many approaches to performing qualitative risk analysis
- Same basic steps as quantitative analysis
- Still identifying assets, threats, vulnerabilities and controls
- Just evaluating importance differently

Example Seven Steps of Risk Analysis

Step 1. Identify scope
Bound the problem.

Step 2. Assemble team
Include subject matter experts, management in charge of implementing, and users.

Step 3. Identify threats
Pick from list of known threats.
Brainstorm new threats.
Mix threats and vulnerabilities.

Step 4. Threat prioritization
Prioritize threats for each asset.
Predict likelihood of occurrence.
Define a fixed threat rating.
(e.g., low (1) .... high (5)
Associate a rating with each threat.
Approximate the risk probability in the quantitative approach.

Step 5. Loss impact

Step 6. Identify controls and safeguards

Step 7. Communicate results

EXERCISES

1. What is an *information system (IS)*? What are its components?

2. How does IS relate to businesses and help them?

3. What are the different types of IS? Discuss in detail *management support systems*.

4. What *different approaches* are followed in developing IS? Explain one of the approaches in brief.

5. What is the meaning of *information security*? Discuss the goals of information security and the relationship between these goals.

6. What is *information assurance (IA)*? Differentiate between *information security (IS)* and *information assurance (IA)*. Also describe the three-dimensional model with respect to *IA*.

7. What is *cybersecurity*? Explain some of the measures taken to ensure cybersecurity.

8. How do you classify information systems in general? Explain in detail.

9. Explain in detail the different approaches followed to develop information systems.

10. What is *information assurance* and how it is provided?

11. Describe *security risk analysis*.

12. Discuss the various goals of *information security*.

13. What are the types of threats in an *information system*?

14. Diagrammatically show the *information assurance model*.

15. Prove this hypothesis – "The Internet is the hope of the future."

16. Mention the important models of *IS development*.

17. "All the information in a computer system is completely secure." State your views for/against this statement.

18. What is meant by *enterprise information system*.

19. Define the term *TPS* and add a note on its advantage.

20. Describe *security risk analysis* in detail.

21. What is the meaning of *system development*? Describe any two models of system development.

22. What are various threats to *information systems*?

23. Differentiate between *security and threats*.

24. Explain the concept of *Web security*.

25. What is the classification of threats? Describe the role of Web services.

26. What are *decision support systems*? Discuss major functional requirement of a decision support system.

27. What are security threats in mobile computing? Explain how to deal with them.

28. Distinguish between *information-level threats and network-level threats*.

29. Explain different roles involved in managing information as "assets." How does the role of custodian differ from that of user?

30. What are the *principles of information security*?

31. Explain the concept of the *CIA triad*.

32. Briefly describe some *important characteristics of information*.

33. What do you understand by *value of information*?

34. Give the *criteria for classification of information*.

35. Define *active and passive attacks*.

36. What is *RFID*? How can RFID be used in mobile commerce and information asset protection?

CHAPTER 2

CYBERSECURITY APPLICATION SECURITY

2.1 APPLICATION SECURITY

Application security is the use of software, hardware, and procedural methods to protect applications from external threats. Once an afterthought in software design, security is becoming an increasingly important concern during development as applications become more frequently accessible over networks and are, as a result, vulnerable to a wide variety of threats. Security measures built into applications and a sound application security routine minimize the likelihood that unauthorized code will be able to manipulate applications to access, steal, modify, or delete sensitive data.

Actions taken to ensure application security are sometimes called countermeasures. The most basic software countermeasure is an application firewall that limits the execution of files or the handling of data by specific installed programs. The most common hardware countermeasure is a router that can prevent the IP address of an individual computer from being directly visible on the Internet. Other countermeasures include conventional firewalls, encryption/decryption programs, anti-virus programs, spyware detection/removal programs, and biometric authentication systems.

Application security can be enhanced by rigorously defining enterprise assets, identifying what each application does (or will do) with respect to these assets, creating a security profile for each application, identifying and prioritizing potential threats, and documenting adverse events and the actions taken in each case. This process is known as threat modeling. In this context, a threat is any potential or actual adverse event that can comprise the assets of an enterprise, including both malicious events, such as denial of service (DoS) attack, and unplanned events, such as failure of a storage device.

2.1.1 Database

Database security is a growing concern evidenced by an increase in the number of reported incidents of loss of unauthorized exposure to sensitive data. As the amount of data collected, retained, and shared electronically expands, so does the need to understand database security.

As its core, database security strives to ensure that only authenticated users perform authorized activities at authorized times. While database security incorporates a wide array of security topics, notwithstanding physical security, network security, encryption, and authentication, this text focuses on the concepts and mechanisms particular to securing data. Within that context, database security encompasses three constructs:

1. Confidentiality or protection of data from unauthorized users

2. Integrity or prevention from unauthorized data access

3. Availability or identification of and recovery from hardware and software errors or malicious activity resulting in denial of data availability

2.1.1.1 Access Control

Generally, access control is defined in three different ways:

a. Mandatory access control (MAC)

b. Discretionary access control (DAC)

c. Role-based access control (RBAC)

MAC and DAC provide privileges to specific users or groups to which users are assigned.

2.1.1.2 Access Control Commands: Grant and Revoke

Access control is a core concept in security. Access control limits actions on objects to specific users. In database security, objects pertain to data objects such as tables and columns as well as SQL objects such as views and stored procedures. Data actions include read (select), insert, update, and delete or execute for stored procedures. For instance, a faculty member, Professor X, may be given read privileges to the student table.

2.1.1.3 Access Control Categorization

Access control is about restricting the access of either the physical area or the logical area of concern for implementing the security measures. Intrusion of unauthorized personnel to the secured areas can be prevented using access control mechanisms. These are basically of two types, based on types of access areas:

1. **Physical access control:** The physical access control mechanism prevents unauthorized users from entering the restricted area and thus secures the physical area from intrusion and number of attacks.

2. **Logical access control:** The logical access control mechanism prevents unauthorized access to logical areas like websites, email accounts, applications, and events. This secures the system against malware programs and also unauthorized users of the system.

2.1.2 Email

In today's electronic world, email is critical to any business being competitive. In most cases it now forms the backbone of most organization's day-to-day activities, and its use will continue to grow. As email becomes more prevalent in the market, the importance of email security becomes more significant, in particular, the security implications associated with the management of email storage, policy enforcement, auditing and archiving, and data recovery. Managing large, active stores of information takes time and effort in order to avoid failures—failures that will impact the users and therefore the business, undoubtedly leading to lost productivity. For secure and effective storage management, organizations must take a proactive approach and invest wisely in a comprehensive solution.

When considering a secure email storage management solution, a layered approach, combining both business processes and applications, make sense, considering the service email provides to business. Email management can be broken down into a number of components:

- Mail flow
- Storage
- User access (both at server and user levels)

While each one of these components should be addressed separately, they must be viewed as part of a total security agenda.

Mail flow can encompass many aspects of an email system. However, the security of mail flow is for the most part focused on the auditing and tracking of mails into and out of the organization. Monitoring the content and ensuring that any email that has been sent and received complies with business policy is fundamental. Proving who has sent or received email is a lawful requirement for many industries, and email can often be used as evidence in fraud and human resource court cases.

Another key aspect of the management of mail flow security is the protection of the business from malicious or unlawful attacks. It is the gateway into the mail system where a business must protect itself via a variety of methods including hardware and software protecting systems, such as spam filters and virus scanners.

Storing actual email data includes physical storage, logical storage, archiving systems, as well as backup and recovery solutions. The biggest security threat to any email storage system is the potential for mail data to be lost. Most organizations see this threat as existing in the data center and spend millions on securing it. In fact, the threat is most likely to come from lost or stolen hardware, such as laptops containing official email files. When we consider that the number of employees working remotely is growing, including those who only work away from the office periodically, email security on laptops becomes more significant. Providing a managed method of archiving and controlling this data is therefore essential.

When it comes to archiving, organizations should take a two-pronged approach to reduce the risk and retain corporate knowledge. First, users should be frequently educated about email retention policies. In addition, an archiving solution should enable administrators to remove items from users' mailboxes based on administrator-configured options such as the age or size of a message. Administrators should be able to control, retain, and back up the email files by consolidating the information stored in email files while ensuring that users are prevented from simply creating new emails.

2.1.3 Internet

Internet security is a tree branch of computer security specifically related to the Internet, often involving browser security but also network security on a more general level as it applies to other applications or operating systems as a whole. Its objective is to establish rules and measures to use against attacks over the Internet. The Internet represents an insecure channel for exchanging information, leading to a high risk of intrusion or fraud, such as phishing. Different methods have been used to protect the transfer of data, including encryption.

Types of Security

1. Network layer security
2. Internet protocol security (IPSec)
3. Security token

1. Network layer security: TCP/IP, which stands for transmission control protocol (TCP) and Internet protocol (IP) (aka Internet protocol suite), can be made secure with the help of cryptographic methods and protocols. These protocols include a secure sockets layer (SSL), succeeded by transport layer security (TLS) for Web traffic, pretty good privacy (PGP) for email, and IPSec for the network layer security.

2. Internet protocol security (IPSec): This protocol is designed to protect communication in a secure manner using TCP/IP (aka Internet protocol suite). It is a set of security extensions developed by the Internet Task Force (IETF), and it provides security and authentication at the IP layer by transforming data using encryption. There are two main types of transformations that form the basis of IPSec:

- Authentication header (AH)
- Encapsulating security payload (ESP)

These two protocols provide data integrity, data origin authentication, and anti–replay service. These protocols can be used alone or in combination to provide the desired set of security services for the Internet protocol (IP) layer.

3. Security token: Some online sites offer customers the ability to use a six-digit code, which randomly changes every thirty to sixty seconds on a security token. The keys on the security token have built-in mathematical computations and manipulate numbers based on the current time built into the device. This means that every thirty seconds there is only a certain array of correct numbers possible to validate access to the online account. The website that the user is logging into would be made aware of that device's serial number and would know the computation and correct time built into the device to verify that the number given is indeed one of the handful of six-digit numbers that works in that given thirty–sixty second cycle. After thirty to sixty seconds the device will present a new random six-digit number that can log into the website.

2.2 DATA SECURITY CONSIDERATIONS: BACKUPS

Backups

A data backup is the result of copying or archiving files and folders for the purpose of being able to restore them in case of data loss.

Data loss can be caused by many things, ranging from computer viruses to hardware failures, to file corruption to fire, or theft. If we are responsible for business data, a loss may involve critical financial, customer, and company data. If the data is on a personal computer, we could lose financial data and other key files, pictures, music, and so on that would be hard to replace.

As part of a data backup plan, we should consider the following:

- What data (files or folders) to backup

- What compression method to use

- How often to run your backups

- What type of backups to run

- The kind of media on which to store the backups

- Where to store the backup data for safekeeping

The frequency to run your backups depends on how often the data changes.

- Business data that changes frequently may need daily or hourly backups.

- Data that changes every few days might use a weekly or even monthly backup.

- For some data, we might run our backup whenever we make a change.

A best practice for storing backups is to keep a copy of the backup files on site (in our home or office) for easy access and a copy of off site in case of fire, flood, or other damage to the location, which could damage or destroy the on-site backup copy.

Backup Obstacles

- Backup window
- Network bandwidth
- System throughput
- Lack of resources

2.3   ARCHIVAL

Data archiving is the process of moving data that is no longer actively used to separate data storage device for long-term retention. Data archives consist of older data that is still important and necessary for future reference, as well as data that must be retained for regulatory compliance. Data archives are indexed and have search capabilities so that files and parts of files can be easily located and retrieved.

Data archives are often confused with data backups, which are copies of data. Data backups are used to restore data in case it is corrupted or destroyed. In contrast, data archives protect older information that is not needed for everyday operations but may occasionally need to be accessed.

In computers, archival storage is storage for data that may not be actively needed but is kept for possible future use or for record-keeping purposes. Archival storage is often provided using the same system as that used for backup storage. Typically, archival and backup storage can be retrieved using a restore process.

2.3.1 Storage and Disposal of Data

In many cases, the single most expensive item in a backup project is the backup storage device itself. Therefore, it is important that the technical specifications of the storage device provide adequate capacity and performance to accommodate existing and planned data. Determining the tape format, number of tape drives, and how many slots are required is predicated on many variables. Backup windows, growth rates, retention policies, duplicate tape copies, and network and server throughputs all affect which backup storage device is best for our needs.

Tape libraries are sized using two variables:

The number of tape drives and the number of slots; manufacturers of tape libraries continue to improve each of these elements.

Tape libraries today are available with five to fifty thousand slots and can support anywhere from one to 256 tape drives. Additionally, there are tape libraries available that support multiple tape formats.

When designing a centralized data backup, take particular care selecting the right backup storage device. Make sure it can easily scale as data rates increase. Verify that the shelf life of the media meets long-term storage needs. Calculate the required throughput to meet the backup window and make sure that you can support enough tape drives to meet this window.

2.3.2 Secure Data Disposal Methods

Information systems store data on a wide variety of storage media, including internal and external hard drives; internal solid-state memory, removable flash memory cards, and flash drives; floppy, ZIP, and other types of removable magnetic disks; tapes, cartridges, and other linear magnetic media; optical storage using CDs and DVDs; and paper.

To prevent unauthorized access, it is critical that data be rendered unreadable when it or the device on which it resides is no longer needed. This is required by law (and common sense) for all computers and media containing sensitive information.

Note that different kinds of data storage media require different methods for secure removal or destruction, some simple but others complex. Do it incorrectly and the data remains for prying eyes to discover.

Proof that secure disposal is not easy comes from this simple fact: Insecure disposal is one of the most common causes of sensitive data being compromised. Not coincidentally, it is one of the most common methods by which identity theft occurs.

2.3.2.1 Paper Media

Paper containing sensitive information should be shredded. Every office (and home) should have access to a shredder or a secure shredding service. Shredders are cheap. "Dumpster diving" for data is common. Secure recycling containers are distributed around the medical campus for just this reason.

Alternatively, paper records can be pulverized (rendered into a powder by grinding), macerated (rendered into pulp by chemicals), or incinerated (burned). This is appropriate for extremely sensitive information.

2.3.2.2 Electronic Media

The appropriate "cleaning" method for electronic media depends on the type. The main division is between "magnetic media" and "optical media." Though both contain information in electronic form, the methods for secure disposal are very different.

Many people are under the impression that all they need to do is "delete" a file from computer hard drive or other storage media. Unfortunately, that's almost never sufficient. In most cases, "delete" simply changes indexing information about a file, sort of like marking through the entry in a book's table of contents but leaving the pages behind.

Emptying the "recycle bin" or the "trash" folder of deleted flies is usually ineffective. These methods remove the pointers (indexes) to the deleted files, but the data itself still remains on the storage media as unallocated space.

Even if the unallocated space is subsequently used by new files, there are sophisticated scanning methods that could be used to recover data previously stored in those locations.

Some un-rewritable media, like CDs and DVDs, cannot have their contents deleted in any case. Inoperable media, like a crashed hard drive, may be so corrupted that we cannot access it using normal computer operations; but it still may have data on it that can be recovered by others.

2.3.2.3 Demagnetizing Magnetic Media

Removable magnetic "disks" (floppies, ZIP disks, and the like) and linear magnetic media (tape reels, cartridges) can be "degaussed"—that is, demagnetized. An appropriately sized and appropriately powered "degausser" is required.

For each particular type of magnetic storage and size of degausser there is a minimum erasing time.

As with disposal of paper information, there are trade-offs rather than absolute standards for "erasing" magnetic media. The more powerful and lengthy the degaussing process applied to any given type of storage media, the less likely it is to be subsequently recovered by others.

Note that degaussing can make the media inoperable, so this method is not recommended if the media needs to be reused and/or has resale value.

2.3.2.4 Over Writing Magnetic Media

"Fixed" internal magnetic storage, such as computer hard drives, as well as external "mini" and "micro" hard drives, can be cleaned by software that uses an overwriting or "wiping" process. USB "flash drives" devices and plug-in memories like compact flashes, memory sticks, and secure digital and smart media can also be cleaned in this way.

Special software is used to overwrite all the useable storage locations. The simplest method is a single overwrite; additional security is provided by multiple overwrites with variations of all 0's and 1's, complements (opposite of recorded characters) and/or random characters so that recovery even by the most sophisticated methods becomes almost impossible.

2.3.2.5 Mangling Magnetic Media

We can take a hammer or a high-speed drill to our hard drive, USB drive, or other device. Chances are excellent that we will render it inoperable in short order.

But be warned that recovery of data from physically mangled magnetic devices is still impossible. Physical destruction is generally something that must be done by a trained person to be completely effective, particularly for hard drives.

Floppy disks can be broken open and the internal magnetic disk cut up. As with optical media, caution is required to avoid personal injury from flying plastic parts, and it is still theoretically possible to recover data even from a mangled disk.

2.3.2.6 Optical Media

"Write many" optical media (such as CD–RWs and DVD–RWs) can be processed via an overwrite method similar to that for magnetic media. However, the vast majority of optical media in use are of the "write once" type—notably the ubiquitous CDs and DVDs. They cannot be overwritten. Because such media are optical rather than magnetic, neither they can't be degaussed either.

So, as with paper, only physical destruction will do. Many higher-capacity paper shredders are rated for CD/DVD destruction for exactly this reason. It is a good investment to upgrade a shredder that is CD/DVD capable if you regularly rely on optical media for our data storage.

As with magnetic media, we can perform a physical attack. Cutting a CD or DVD with scissors is an alternative if there are only a few. But note that cut–up discs have been successfully reassembled and read, so cut them into multiple pieces and, ideally, dispose of the pieces in different trash receptacles.

Breaking discs in half with our hands can send dangerous shards of plastic flying. Burning discs (or microwaving them) can release toxic fumes.

2.3.2.7 Computer Recycling Programs

For a whole system, some manufactures like Dell and Apple, and many retailers of computer equipment, offer recycling programs that meet both security and environmental concerns. These programs will process the entire old system for disposal, including cleaning the hard drive and any other storage media, when we trade it in as part of a new purchase.

2.4 SECURITY TECHNOLOGY

2.4.1 Firewall

A computer firewall controls access between networks. It generally consists of gateways and filters, which vary from one firewall to another. Firewalls also screen network traffic and are able to block traffic that is dangerous. Firewalls act as the intermediate server between simple mail transfer protocol (SMTP) and hypertext transfer protocol (HTTP) connections.

Firewalls impose restrictions on incoming and outgoing network packets to and from private networks. Incoming or outgoing traffic must pass through the firewall; only authorized traffic is allowed to pass through it. Firewalls create check points between an internal private network and the public Internet, also known as choke points (borrowed from the identical military term of a combat-limiting geographical feature). Firewalls can create choke points based on IP source and TCP port number. They can also serve as the platform for IPSec. Using tunnel mode capability, firewalls can be used to implement VPNs. Firewalls can also limit network exposure by hiding the internal network system and information from the public Internet.

Types of Firewalls

1. Packet filter

2. Stateful packet inspection

3. Application-level gateway

1. Packet filter: A packet filter is a first-generation firewall that processes network traffic on a packet–by–packet basis. Its main job is to filter traffic from a remote IP host, so a router is needed to connect the internal network to the Internet. The router is known as a screening router, which screens packets leaving and entering the network.

2. Stateful packet inspection: In a stateful firewall the circuit-level gateway is a proxy server that operates at the network level of an open system interconnection (OSI) model and statically defines what traffic will be allowed. Circuit proxies will forward network packets (formatted unit of data) containing a given port number, if the port is permitted by the algorithm. The main advantage of a proxy server is its ability to provide network address translation (NAT), which can hide the user's IP address from the Internet, effectively protecting all internal information from the Internet.

3. Application–level gateway: An application–level firewall is a third-generation firewall where a proxy server operates at the very top of the open system interconnection (OSI) model, the IP suite application level. A network packet is forwarded only if a connection is established using a known protocol. Application–level gateways are notable for analyzing entire messages rather than individual packets of data when the data are being sent or received.

2.4.2 Virtual Private Networks (VPNs)

VPN is pronounced as separate letters and is short for virtual private network. VPN is a network that is constructed by using public wires—usually the Internet—to connect to a private network, such as company's internal network. There are a number of systems that enable us to create networks using the Internet as the medium for transporting data. These systems use encryption and other security mechanisms to ensure that only authorized users can access the network and that the data cannot be intercepted.

Consumer VPN Service

Consumers use a private VPN service, also known as a VPN tunnel, to protect their online activity and identity. By using an anonymous VPN service, a user's Internet traffic and data remain encrypted, which prevents eavesdroppers from sniffing Internet activity. A VPN service is especially useful when accessing public Wi-Fi hotspots because the public wireless services might not be secure. In addition to public Wi-Fi security, a private VPN service also provides consumers with uncensored Internet access and can help prevent data theft and unblock websites.

Corporate VPN Communications

Companies and organizations will use a VPN to communicate confidentially over a public network and to send voice, video, or data. It is also an excellent option for remote workers and organizations with global offices and partners to share data in a private manner.

2.5   INTRUSION DETECTION

Intrusion detection is exactly what is sounds like. It is the act of identifying intruders on a network. This can be done at many different levels:

- At the network (point of entry) firewall
- At any point in the network
- At a specific host

This is usually done by matching signatures (or patterns) of known attacks against network traffic.

An *intrusion detection system (IDS)* is a device or software application that monitors network or system activities for malicious activities or policy violations and produces electronic reports to a management station. IDSs come in a variety of "flavors" and approach the goal of detecting suspicious traffic in different ways. There are network-based (NIDS) and host-based (HIDS) intrusion detection systems. NIDS is a network security system focusing on the attacks that come from the inside of the network (authorized users). Some systems may attempt to stop an intrusion attempt but this is neither required nor expected of a monitoring system. Intrusion detection and prevention systems (IDPS) are primarily focused on identifying possible incidents, logging information about them, and reporting attempts. In addition, organizations use IDPSes for other purposes, such as identifying problems with security policies, documenting existing threats, and deterring individuals from violating security policies. IDPSs have become a necessary addition to the security infrastructure of nearly every organization.

IDPSs typically record information related to observed events, notify security administrators of important observed events, and produce reports. Many IDPSs can also respond to a detected threat by attempting to prevent it from succeeding. They use several response techniques, which involve the IDPS stopping the attack itself, changing the security environment (e.g., reconfiguring a firewall), or changing the attack's content.

Terminology

Alarm filtering: The process of categorizing attack alerts produced from an IDS in order to distinguish false positives from actual attacks.

Attacker or **intruder:** An entity that tries to find a way to gain unauthorized access to information, inflict harm, or engage in other malicious activities.

Burglar alarm: A signal suggesting that a system has been or is being attacked.

Clandestine user: A person who acts as a supervisor and tries to use his privileges so as to avoid being captured.

Confidence value: A value an organization places on an IDS based on past performance and analysis to help determine its ability to effectively identify an attack.

Detection rate: The detection rate is defined as the number of intrusion instances detected by the system (true positive) divided by the total number of intrusion instances present in the test set.

False alarm rate: Defined as the number of "normal" patterns classified as attacks (false positive) divided by the total number of "normal" patterns.

False negative: When no alarm is raised when an attack has taken place.

False positive: An event signaling an IDS to produce an alarm when no attack has taken place.

Masquerader: A person who attempts to gain unauthorized access to a system by pretending to be an authorized user. They are generally outside users.

Misfeasor: They are commonly internal users and can be of two types:

1. An authorized user with limited permissions

2. A user with full permissions who misuses their powers

Noise: Data or interference that can trigger a false positive or obscure a true positive.

Site policy: Guidelines within an organization that control the rules and configurations of an IDS.

Site policy awareness: An IDS's ability to dynamically change its rules and configurations in response to changing environmental activity.

True negative: An event when no attack has taken place and no detection is made.

True positive: A legitimate attack that triggers an IDS to produce an alarm.

2.5.1 HIDS and NIDS

Intrusion detection systems are of two main types, network-based (NIDS) and host-based (HIDS) intrusion detection systems.

Network Intrusion Detection Systems

Network intrusion detection systems (NIDS) are placed at a strategic point or points within the network to monitor traffic to and from all devices on the network. It performs an analysis of passing traffic on the entire subnet and matches the traffic that is passed on the subnets to the library of known attacks. Once an attack is identified, or abnormal behavior is sensed, the alert can be sent to the administrator. An example of an NIDS would be installing it on the subnet where firewalls are located in order to see if someone is trying to break into the firewall. Ideally one would scan all inbound and outbound traffic; however, doing so might create a bottleneck that would impair the overall speed of the network. OPNET and NetSim are commonly used tools for simulation network intrusion detection systems. NID systems are also capable of comparing signatures for similar packets to link and drop harmful detected packets, which have a signature matching the records in the NIDS. When we classify the designing of the NIDS according to the system interactivity property, there are two types: online and off-line NIDS. Online NIDS deals with the network in real time. It analyses the ethernet packets and applies some rules to decide if it is an attack or not. Off-line NIDS deals with stored data and passes it through some processes to decide if it is an attack or not.

Host Intrusion Detection Systems

Host intrusion detection systems (HIDS) run on individual hosts or devices on the network. A HIDS monitors the inbound and outbound packets from the device only and will alert the user or administrator if suspicious activity is detected. It takes a snapshot of existing system files and matches it to the previous snapshot. If the critical system files were modified or deleted, an alert is sent to the administrator to investigate. An example of HIDS usage can be seen on mission critical machines, which are not expected to change their configurations.

Intrusion detection systems can also be system specific using custom tools and honeypots.

Passive and Reactive Systems

In a passive system, the intrusion detection system (IDS) sensor detects a potential security breach, logs the information, and signals an alert on the console or owner. In a reactive system, also known as an intrusion prevention system (IPS), the IPS auto-responds to the suspicious activity by resetting the connection or by reprogramming the firewall to block network traffic from the suspected malicious source. The term "IDPS" is commonly used where this can happen automatically or at the command of an operator: systems that both "detect (alert)" and "prevent."

Comparison with Firewalls

Though they both relate to network security, an intrusion detection system (IDS) differs from a firewall in that a firewall looks outwardly for intrusions in order to stop them from happening. Firewalls limit access between networks to prevent intrusion and do not signal an attack from inside the network. An IDS evaluates a suspected intrusion once it has taken place and signals an alarm. An IDS also watches for attacks that originate from within a system. This is traditionally achieved by examining network communications, identifying heuristics and patterns (often known as signatures) of common computer attacks, and taking action to alert operators. A system that terminates connections is called an intrusion prevention system and is another form of an application layer firewall.

2.5.2 Statistical Anomaly and Signature-Based IDSEs

All intrusion detection systems use one of two detection techniques:

Statistical Anomaly-Based IDS

An IDS that is anomaly based will monitor network traffic and compare it against an established baseline. The baseline will identify what is "normal" for that network—what sort of bandwidth is generally used and what protocols are used that it may raise a false positive alarm for a legitimate use of bandwidth if the baselines are not intelligently configured.

Signature-Based IDS

A signature-based IDS will monitor packets on the network and compare them against a database of signatures or attributes from known malicious threats. This is similar to the way most antivirus software detects malware. The issue is that there will be a lag between a new threat being discovered in the wild and the signature for detecting that threat being applied to the IDS. During that lag time the IDS would be unable to detect the new threat.

Limitations

- Noise can severely limit an intrusion detection system's effectiveness. Bad packets generated from software bugs, corrupt DNS data, and local packets that escaped can create a significantly high false-alarm rate.

- It is not uncommon for the number of real attacks to be far below the number of false alarms. The number of real attacks is often so far below the number of false alarms that the real attacks are often missed and ignored.

- Many attacks are geared for specific versions of software that are usually outdated. A constantly changing library of signatures is needed to mitigate threats. Outdated signature databases can leave the IDS vulnerable to newer strategies.

- For signature-based IDSs there will be lag between a new threat discovery and its signature being applied to the IDS. During this lag time the IDS will be unable to identify the threat.

- It cannot compensate for a weak identification and authentication mechanisms or for weaknesses in network protocols. When an attacker gains access due to weak authentication mechanisms, then the IDS cannot prevent the adversary from any malpractice.

- Encrypted packets are not processed by the intrusion detection software. Therefore, the encrypted packet can allow an intrusion to the network that is undiscovered until more significant network intrusions have occurred.

- Intrusion detection software provides information based on the network address that is associated with the IP packet that is sent into the network. This is beneficial if the network address contained in the IP packet is accurate. However, the address that is contained in the IP packet could be faked or scrambled.

- Due to the nature of NIDS systems, and the need for them to analyze protocols as they are captured, NIDS systems can be susceptible to the same protocol-based attacks network hosts may be vulnerable to. Invalid data and TCP/IP stack attacks may cause an NIDS to crash.

Evasion Techniques

There are a number of techniques attackers are using; the following are considered "simple" measures that can be taken to evade IDS:

- **Fragmentation:** By sending fragmented packets, the attacker will be under the radar and can easily bypass the detection system's ability to detect the attack signature.

- **Avoiding defaults:** The TCP port utilized by a protocol does not always provide an indication to the protocol that is being transported. For example, an IDS may expect to detect a Trojan on port 12345. If an attacker had reconfigured it to use a different port, the IDS may not be able to detect the presence of the Trojan.

- **Coordinated, low-bandwidth attacks:** Coordinating a scan among numerous attackers (or agents) and allocating different ports or hosts to different attackers makes it difficult for the IDS to correlate the captured packets and deduce that a network scan is in progress.

- **Address spoofing/proxying:** Attackers can increase the difficulty of the ability of security administrators to determine the source of the attack by using poorly secured or incorrectly configured proxy servers to bounce an attack. If the source is spoofed and bounced by a server, then it makes it very difficult for IDS to detect the origin of the attack.

- **Pattern-change evasion:** IDSs generally rely on "pattern matching" to detect an attack. By changing the data used in the attack slightly, it may be possible to evade detection. For example, an IMAP server may be vulnerable to a buffer overflow, and an IDS is able to detect the attack signature of ten common attack tools. By modifying the payload sent by the tool so that it does not resemble the data that the IDS expects, it may be possible to evade detection.

Development

One preliminary IDS concept consists of a set of tools intended to help administrators review audit trails. User access logs, file access logs, and system event logs are examples of audit trails.

Fred Cohen noted in 1984 that it is impossible to detect an intrusion in every case, and that the resources needed to detect intrusions grow with the amount of usage.

Dorothy E. Denning, assisted by Peter G. Neumann, published a model of an IDS in 1986 that formed the basis for many systems today. Her model used statistics for anomaly detection and resulted in an early IDS at SRI International named the Intrusion Detection Expert System (IDES), which ran on Sun workstations and could consider both user- and network-level data. IDES had a dual approach with a rule-based expert system to detect known types of intrusions plus a statistical anomaly detection component based on profiles of users, host systems, and target systems. Lunt proposed adding an artificial neural network as a third component. She said all three components could then report to a resolver. SRI followed IDES in 1993 with the Next-Generation Intrusion Detection Expert System (NIDES).

The Multics Intrusion Detection and Alerting System (MIDAS), an expert system using P-BEST and Lisp, was developed in 1988 based on the work of Denning and Neumann. Haystack was also developed in that year using statistics to reduce audit trails.

Wisdom and Sense (W&S) was a statistics-based anomaly detector developed in 1989 at the Los Alamos National Laboratory. W&S created rules based on statistical analysis and then used those rules for anomaly detection.

In 1990, the Time-Based Inductive Machine (TIM) did anomaly detection using inductive learning of sequential user patterns in common Lisp on a VAX 3500 computer. The Network Security Monitor (NSM) performed masking on access matrices for anomaly detection on a Sun-3/50 workstation.

The Information Security Officer's Assistant (ISOA) was a 1990 prototype that considered a variety of strategies including statistics, a profile checker, and an expert system. ComputerWatch at AT&T Bell Labs used statistics and rules for audit data reduction and intrusion detection.

Then, in 1991, researchers at the University of California, Davis created a prototype Distributed Intrusion Detection System (DIDS), which was also an expert system. The Network Anomaly Detection and Intrusion Reporter (NADIR), also in 1991, was a prototype IDS developed at the Los Alamos National Laboratory's Integrated Computing Network (ICN) and was heavily influenced by the work of Denning and Lunt. NADIR used a statistics-based anomaly detector and an expert system.

The Lawrence Berkeley National Laboratory announced Bro in 1998, which used its own rule language for packet analysis from libpcap data. The Network Flight Recorder (NFR) in 1999 also used libpcap. APE was developed as a packet sniffer, also using libpcap, in November 1998, and was renamed Snort one month later. APE has since become the world's largest used IDS/IPS system with over three hundred thousand active users.

The Audit Data Analysis and Mining (ADAM) IDS in 2001 used tcpdump to build profiles of rules for classifications.

In 2003, Dr. Yongguang Zhang and Dr. Wenke Lee argue for the importance of IDS in networks with mobile nodes.

2.6 DENIAL-OF-SERVICE (DOS) ATTACK

A denial-of-service attack (DoS attack) or distributed denial of service attack (DDoS attack) is an attempt to make a computer resource unavailable to its intended users. Although the means to carry out, motives for, and targets of a DoS attack may vary, it generally consists of the concerted efforts to prevent an Internet site or service from functioning efficiently or at all, temporarily or indefinitely.

In computing, a *denial-of-service (DoS) attack* is an attempt to make a machine or network resource unavailable to its intended users, such as to temporarily or indefinitely interrupt or suspend services of a host connected to the Internet. A *distributed denial-of-service (DDoS)* attack is where the attack source is more than one, often thousands of, unique IP addresses. It is analogous to a group of people crowding the entry door or gate to a shop or business and not letting legitimate parties enter the shop or business, disrupting normal operations.

Criminal perpetrators of DoS attacks often target sites or services hosted on high-profile Web servers such as banks and credit card payment gateways; but motives of revenge, blackmail, or activism can be behind other attacks.

Symptoms Exhaust

The United States Computer Emergency Readiness Team (US-CERT) defines symptoms of denial-of-service attacks to include the following:

- Unusually slow network performance (opening files or accessing websites)
- Unavailability of a particular website
- Inability to access any website
- Dramatic increase in the number of spam emails received—(this type of DoS attack is considered an email bomb)
- Disconnection of a wireless or wired Internet connection
- Long-term denial of access to the Web or any internet services

If the attack is conducted on a sufficiently large scale, entire geographical regions of Internet connectivity can be compromised without the attacker's knowledge or intent by incorrectly configured or flimsy network infrastructure equipment.

Attack Techniques

A denial-of-service attack is characterized by an explicit attempt by attackers to prevent legitimate users of a service from using that service. There are two general forms of DoS attacks: those that crash services and those that flood services.

The most serious attacks are distributed and in many or most cases involve forging of IP sender addresses (IP address spoofing) so that the location of the attacking machines cannot easily be identified and filtering cannot be done based on the source address.

Internet Control Message Protocol (ICMP) Flood

A smurf attack relies on misconfigured network devices that allow packets to be sent to all computer hosts on a particular network via the broadcast address of the network rather than a specific machine. The attacker will send large numbers of IP packets with the source address faked to appear to be the address of the victim. The network's bandwidth is quickly used up, preventing legitimate packets from getting through to their destination. They are normally used against small clients or servers, without ISP security.

Ping flood is based on sending the victim an overwhelming number of ping packets, usually using the "ping" command from Unix-like hosts (the -t flag on Windows systems is much less capable of overwhelming a target; also the –l (size) flag does not allow sent packet size greater than 65,500 in Windows). It is very simple to launch, the primary requirement being access to greater bandwidth than the victim.

Ping of death is based on sending the victim a malformed ping packet, which will lead to a system crash on a vulnerable system.

(S) SYN Flood

A SYN flood occurs when a host sends a flood of TCP/SYN packets, often with a forged sender address. Each of these packets are handled like a connection request, causing the server to spawn a half-open connection by sending back a TCP/SYN-ACK packet (acknowledge) and waiting for a packet in response from the sender address (response to the ACK packet). However, because the sender address is forged, the response never comes. These half-open connections saturate the number of available connections the server can make, keeping it from responding to legitimate requests until after the attack ends.

Teardrop Attacks

A teardrop attack involves sending mangled IP fragments with overlapping, oversized payloads to the target machine. This can crash various operating systems because of a bug in their TCP/IP fragmentation re-assembly code. Windows 3.1x, Windows 95, and Windows NT operating systems, as well as versions of Linux prior to versions 2.0.32 and 2.1.63, are vulnerable to this attack.

(Although in September 2009, a vulnerability in Windows Vista was referred to as a "teardrop attack," this targeted SMB2 is a higher layer than the TCP packets that teardrop used.)

Peer-to-Peer Attacks

Attackers have found a way to exploit a number of bugs in peer-to-peer servers to initiate DDoS attacks. The most aggressive of these peer-to-peer DDoS attacks exploits DC++. With peer-to-peer attacks there is no botnet and the attacker does not have to communicate with the clients it subverts. Instead, the attacker acts as a "puppet master," instructing clients of large peer-to-peer file-sharing hubs to disconnect from their peer- to-peer network and to connect to the victim's website instead.

Permanent Denial-of-Service Attacks

Permanent denial-of-service (PDoS) attacks, also known loosely as phlashing,[14] is an attack that damages a system so badly that it requires replacement or reinstallation of hardware. Unlike the distributed denial-of-service attack, a PDoS attack exploits security flaws, which allow remote administration on the management interfaces of the victim's hardware, such as routers, printers, or other networking hardware. The attacker uses these vulnerabilities to replace a device's firmware with a modified, corrupt, or defective firmware image— a process that when done legitimately is known as flashing. This therefore "bricks" the device, rendering it unusable for its original purpose until it can be repaired or replaced.

The PDoS is a pure hardware targeted attack that can be much faster and requires fewer resources than using a botnet or a root/vserver in a DDoS attack. Because of these features, and the potential high probability of security exploits on network-enabled embedded devices (NEEDs), this technique has come to the attention of numerous hacking communities.

PhlashDance is a tool created by Rich Smith (an employee of Hewlett-Packard's Systems Security Lab) used to detect and demonstrate PDoS vulnerabilities at the 2008 EUSecWest Applied Security Conference in London.

Application-Layer Floods

Various DoS-causing exploits such as buffer overflow can cause server-running software to get confused and fill the disk space or consume all available memory or CPU time.

Other kinds of DoS rely primarily on brute force, flooding the target with an overwhelming flux of packets, oversaturating its connection bandwidth or depleting the target's system resources. Bandwidth-saturating floods rely on the attacker having higher bandwidth available than the victim; a common way of achieving this today is via distributed denial of service, employing a botnet. Another target of DDoS attacks may be to produce added costs for the application operator, when the latter uses resources based on cloud computing. In this case normally application-used resources are tied to a needed quality-of-service level (e.g., responses should be less than 200 ms), and this rule is usually linked to automated software (e.g., Amazon CloudWatch) to raise more virtual resources from the provider in order to meet the defined QoS levels for the increased requests. The main incentive behind such attacks may be to drive the application owner to raise the elasticity levels in order to handle the increased application traffic, in order to cause financial losses or force them to become less competitive. Other floods may use specific packet types or connection requests to saturate finite resources by, for example, occupying the maximum number of open connections or filling the victim's disk space with logs.

A "banana attack" is another particular type of DoS. It involves redirecting outgoing messages from the client back onto the client, preventing outside access as well as flooding the client with the sent packets. A LAND attack is of this type.

An attacker with shell-level access to a victim's computer may slow it until it is unusable or crash it by using a fork bomb.

A kind of application-level DoS attack is XDoS (or XML DoS), which can be controlled by modern Web application firewalls (WAFs).

Nuke

A nuke is an old denial-of-service attack against computer networks consisting of fragmented or otherwise invalid ICMP packets sent to the target, achieved by using a modified ping utility to repeatedly send this corrupt data, thus slowing down the affected computer until it comes to a complete stop.

A specific example of a nuke attack that gained some prominence is the WinNuke, which exploited the vulnerability in the NetBIOS handler in Windows 95. A string of out-of-band data was sent to TCP port 139 of the victim's machine, causing it to lock up and display a blue screen of death (BSOD).

HTTP POST DoS Attack

First discovered in 2009, the HTTP POST attack sends a complete, legitimate HTTP POST header, which includes a "content-length" field to specify the size of the message body to follow. However, the attacker then proceeds to send the actual message body at an extremely slow rate (e.g., 1 byte/110 seconds). Due to the entire message being correct and complete, the target server will attempt to obey the content-length field in the header and wait for the entire body of the message to be transmitted, which can take a very long time. The attacker establishes hundreds or even thousands of such connections, until all resources for incoming connections on the server (the victim) are used up, hence making any further (including legitimate) connections impossible until all data has been sent. It is notable that unlike many other (D)DoS attacks, which try to subdue the server by overloading its network or CPU, a HTTP POST attack targets the logical resources of the victim, which means the victim would still have enough network bandwidth and processing power to operate. Further combined with the fact that Apache will, by default, accept requests up to 2 GB in size, this attack can be particularly powerful. HTTP POST attacks are difficult to differentiate from legitimate connections and are therefore able to bypass some protection systems. OWASP, an open source Web application security project, has released a testing tool to test the security of servers against this type of attack.

R-U-Dead-Yet? (RUDY)

This attack targets Web applications by starvation of available sessions on the Web server. Much like Slowloris, RUDY keeps sessions at halt using never-ending POST transmissions and sending an arbitrarily large content-length header value.

Slow-Read Attack

A slow-read attack sends legitimate application-layer requests but reads responses very slowly, thus trying to exhaust the server's connection pool. Slow reading is achieved by advertising a very small number for the TCP "Receive Window" size and at the same time by emptying clients' TCP "Receive" buffer slowly. That naturally ensures a very low data flow rate.

Distributed Attack

A distributed denial-of-service (DDoS) attack occurs when multiple systems flood the bandwidth or resources of a targeted system, usually one or more Web servers. Such an attack is often the result of multiple compromised systems (for example a botnet) flooding the targeted system with traffic. A botnet is a network of zombie computers programmed to receive commands without the owners' knowledge. When a server is overloaded with connections, new connections can no longer be accepted. The major advantages to an attacker of using a distributed denial-of-service attack are that multiple machines can generate more attack traffic than one machine, multiple attack machines are harder to turn off than one attack machine, and the behavior of each attack machine can be stealthier, making it harder to track and shut down. These attacker advantages cause challenges for defense mechanisms. For example, merely purchasing more incoming bandwidth than the current volume of the attack might not help, because the attacker might be able to simply add more attack machines. This, after all, will end up completely crashing a website for periods of time.

Malware can carry DDoS attack mechanisms; one of the better-known examples of this was MyDoom. Its DoS mechanism was triggered on a specific date and time. This type of DDoS involved hard coding the target IP address prior to release of the malware, and no further interaction was necessary to launch the attack.

A system may also be compromised with a Trojan, allowing the attacker to download a zombie agent, or the Trojan may contain one. Attackers can also break into systems using automated tools that exploit flaws in programs that listen for connections from remote hosts. This scenario primarily concerns systems acting as servers on the Web. Stacheldraht is a classic example of a DDoS tool. It utilizes a layered structure where the attacker uses a client program to connect to handlers, which are compromised systems that issue commands to the zombie agents, which in turn facilitate the DDoS attack. Agents are compromised via the handlers by the attacker, using automated routines to exploit vulnerabilities in programs that accept remote connections running on the targeted remote hosts. Each handler can control up to a thousand agents. In some cases a machine may become part of a DDoS attack with the owner's consent, for example, in Operation Payback, organized by the group Anonymous. These attacks can use different types of internet packets such as TCP, UDP, ICMP, and so on.

These collections of system compromisers are known as botnets/rootservers. DDoS tools like Stacheldraht still use classic DoS attack methods centered on IP spoofing and amplification like smurf attacks and fraggle attacks (these are also known as bandwidth consumption attacks). SYN floods (also known as resource starvation attacks) may also be used. Newer tools can use DNS servers for DoS purposes. Unlike MyDoom's DDoS mechanism, botnets can be turned against any IP address. Script kiddies use them to deny the availability of well-known websites to legitimate users. More sophisticated attackers use DDoS tools for the purposes of extortion—even against their business rivals.

Simple attacks such as SYN floods may appear with a wide range of source IP addresses, giving the appearance of a well-distributed DoS. These flood attacks do not require completion of the TCP three-way handshake and attempt to exhaust the destination SYN queue or the server bandwidth. Because the source IP addresses can be trivially spoofed, an attack could come from a limited set of sources, or may even originate from a single host. Stack enhancements such as syn cookies may be effective mitigation against SYN queue flooding; however, complete bandwidth exhaustion may require involvement.

If an attacker mounts an attack from a single host it would be classified as a DoS attack. In fact, any attack against availability would be classed as a denial-of-service attack. On the other hand, if an attacker uses many systems to simultaneously launch attacks against a remote host, this would be classified as a DDoS attack.

UK's GCHQ has tools built for DDoS, named PREDATORS FACE and ROLLING THUNDER.

Reflected/Spoofed Attack

A distributed denial-of-service attack may involve sending forged requests of some type to a very large number of computers that will reply to the requests. Using Internet Protocol address spoofing, the source address is set to that of the targeted victim, which means all the replies will go to (and flood) the target.

ICMP echo request attacks (smurf attacks) can be considered one form of reflected attack, as the flooding host(s) send echo requests to the broadcast addresses of mis-configured networks, thereby enticing hosts to send echo reply packets to the victim. Some early DDoS programs implemented a distributed form of this attack.

Many services can be exploited to act as reflectors, some harder to block than others. US-CERTs have observed that different services imply different amplification factors, as you will see later.

DNS amplification attacks involve a new mechanism that increased the amplification effect, using a much larger list of DNS servers than seen earlier. SNMP and NTP can also be exploited as reflectors in an amplification attack.

Telephony Denial-of-Service (TDoS)

Voice over IP has made abusive origination of large numbers of telephone voice calls inexpensive and readily automated while permitting call origins to be misrepresented through caller ID spoofing.

According to the US Federal Bureau of Investigation, telephony denial of service (TDoS) has appeared as part of various fraudulent schemes:

- A scammer contacts the victim's banker or broker, impersonating the victim to request a funds transfer. The banker's attempt to contact the victim for verification of the transfer fails as the victim's telephone lines are being flooded with thousands of bogus calls, rendering the victim unreachable.

- A scammer contacts consumers with a bogus claim to collect an outstanding payday loan for thousands of dollars. When the consumer objects, the scammer retaliates by flooding the victim's employer with thousands of automated calls. In some cases, displayed caller ID is spoofed to impersonate police or law enforcement agencies.

- A scammer contacts consumers with a bogus debt collection demand and threatens to send police; when the victim balks, the scammer floods local police numbers with calls on which caller ID is spoofed to display the victim's number. Police soon arrive at the victim's residence, attempting to find the origin of the calls.

Telephony denial of service can exist even without Internet telephony. In the 2002 New Hampshire Senate election phone jamming scandal, telemarketers were used to flood political opponents with spurious calls to jam phone banks on election day. Widespread publication of a number can also flood it with enough calls to render it unusable, as happened with multiple +1–area code 867-5309, where subscribers were inundated by hundreds of misdialed calls daily in response to a popular song.

TDoS differs from other telephone harassment (such as prank calls and obscene phone calls) by the number of calls originated; by occupying lines continuously with repeated automated calls, the victim is prevented from making or receiving both routine and emergency telephone calls.

Related exploits include SMS flooding attacks and black fax or fax loop transmission.

Sophisticated Low-Bandwidth Distributed Denial-of-Service Attack

A sophisticated low-bandwidth DDoS attack is a form of DoS that uses less traffic and increases its effectiveness by aiming at a weak point in the victim's system design (i.e., the attacker sends traffic consisting of complicated requests to the system). Essentially, a sophisticated DDoS attack is lower in cost due to its use of less traffic and is smaller in size, making it more difficult to identify, and it has the ability to hurt systems that are protected by flow-control mechanisms.

Denial-of-Service Level II

The goal of a DoS L2 (possibly DDoS) attack is to cause a launching of a defense mechanism that blocks the network segment from which the attack originated. In the case of distributed attack or IP header modification (that depends on the kind of security behavior) it will fully block the attacked network from the Internet, but without a system crash.

Advanced Persistent DoS (APDoS)

An APDoS is more likely to be perpetrated by an advanced persistent threat (APT): actors who are well resourced and exceptionally skilled and have access to substantial commercial-grade computer resources and capacity. APDoS attacks represent a clear and emerging threat needing specialized monitoring and incident response services and the defensive capabilities of specialized DDoS mitigation service providers. This type of attack involves massive network layer DDoS attacks through to focused application layer (HTTP) floods, followed by repeated (at varying intervals) SQLI and XSS attacks. Typically, the perpetrators can simultaneously use from two to five attack vectors involving up to several tens of millions of requests per second, often accompanied by large SYN floods that can not only attack the victim but also any service provider implementing any sort of managed DDoS mitigation capability. These attacks can persist for several weeks—the longest continuous period noted so far lasted 38 days. This APDoS attack involved approximately 50-plus petabits (51,000-plus terabits) of malicious traffic. Attackers in this scenario may (or often will) tactically switch between several targets to create a diversion to evade defensive DDoS countermeasures but all the while eventually concentrating the main thrust of the attack onto a single victim. In this scenario, threat actors with continuous access to several very powerful network resources are capable of sustaining a prolonged campaign, generating enormous levels of un-amplified DDoS traffic.

APDoS Attacks Characterizations

- Advanced reconnaissance (pre-attack OSINT and extensive decoyed scanning crafted to evade detection over long periods)

- Tactical execution (attack with a primary and secondary victims but focus on primary)

- Explicit motivation (a calculated end game/goal target)

- Large computing capacity (access to substantial computer power and network bandwidth resources)

- Simultaneous multi-threaded ISO layer attacks (sophisticated tools operating at layers 3 through 7)

- Persistence over extended periods

DDoS Extortion

In 2015, DDoS botnets such as DD4BC grew in prominence, taking aim at financial institutions. Cyber-extortionists typically begin with a low-level attack and a warning that a larger attack will be carried out if a ransom is not paid in Bitcoin. Security experts recommend targeted websites to not pay the ransom. The attackers tend to get into an extended extortion scheme once they recognize that the target is ready to pay.

Attack Tools

A wide array of programs are used to launch DoS-attacks.

In cases such as MyDoom the tools are embedded in malware, and attacks are launched without the knowledge of the system owner. Stacheldraht is a classic example of a DDoS tool. It utilizes a layered structure where the attacker uses a client program to connect to handlers, which are compromised systems that issue commands to the zombie agents, which in turn facilitate the DDoS attack. Agents are compromised via the handlers by the attacker, using automated routines to exploit vulnerabilities in programs that accept remote connections running on the targeted remote hosts. Each handler can control up to a thousand agents.

In other cases a machine may become part of a DDoS attack with the owner's consent, for example, in Operation Payback, organized by the group Anonymous. The LOIC has typically been used in this way. Along with HOIC, a wide variety of DDoS tools are available today, including paid and free versions, with different features available. There is an underground market for these in hacker-related forums and IRC channels.

UK's GCHQ has tools built for DDoS, named PREDATORS FACE and ROLLING THUNDER.

Defense Techniques

Defensive responses to denial-of-service attacks typically involve the use of a combination of attack detection, traffic classification, and response tools, aiming to block traffic that they identify as illegitimate and allow traffic that they identify as legitimate.

A list of prevention and response tools is provided below:

Firewalls

In the case of a simple attack, a firewall could have a simple rule added to deny all incoming traffic from the attackers, based on protocols, ports, or the originating IP addresses.

More complex attacks will, however, be hard to block with simple rules; for example, if there is an ongoing attack on port 80 (Web service), it is not possible to drop all incoming traffic on this port because doing so will prevent the server from serving legitimate traffic. Additionally, firewalls may be too deep in the network hierarchy, with routers being adversely affected before the traffic gets to the firewall.

Switches

Most switches have some rate-limiting and ACL capability. Some switches provide automatic and/or system-wide rate limiting, traffic shaping, delayed binding (TCP splicing), deep packet inspection, and bogon filtering (bogus IP filtering) to detect and remediate denial-of-service attacks through automatic rate filtering and WAN link failover and balancing.

These schemes will work as long as the DoS attacks can be prevented by using them. For example, SYN flood can be prevented using delayed binding or TCP splicing. Similarly, content-based DoS may be prevented using deep packet inspection. Attacks originating from dark addresses or going to dark addresses can be prevented using bogon filtering. Automatic rate filtering can work as long as set rate thresholds have been set correctly and granularly. Wan-link failover will work as long as both links have DoS/DDoS prevention mechanisms.

Routers

Similar to switches, routers have some rate-limiting and ACL capability. They, too, are manually set. Most routers can be easily overwhelmed under a DoS attack. Cisco IOS has optional features that can reduce the impact of flooding.

Application Front-End Hardware

Application front-end hardware is intelligent hardware placed on the network before traffic reaches the servers. It can be used on networks in conjunction with routers and switches. Application front-end hardware analyzes data packets as they enter the system and then identifies them as priority, regular, or dangerous. There are more than 25 bandwidth management vendors.

Application-Level Key Completion Indicators

In order to meet the case of application-level DDoS attacks against cloud-based applications, approaches may be based on an application-layer analysis to indicate whether an incoming traffic bulk is legitimate and thus enable the triggering of elasticity decisions without the economical implications of a DDoS attack. These approaches mainly rely on an identified path of value inside the application and monitor the macroscopic progress of the requests in this path, toward the final generation of profit, through markers denoted as key completion indicators.

IPS-Based Prevention

Intrusion-prevention systems (IPS) are effective if the attacks have signatures associated with them. However, the trend among the attacks is to have legitimate content but bad intent. Intrusion-prevention systems that work on content recognition cannot block behavior-based DoS attacks.

An ASIC-based IPS may detect and block denial-of-service attacks because they have the processing power and the granularity to analyze the attacks and act like a circuit breaker in an automated way.

A rate-based IPS (RBIPS) must analyze traffic granularly and continuously monitor the traffic pattern and determine if there is traffic anomaly. It must let the legitimate traffic flow while blocking the DoS attack traffic.

DDS-Based Defense

More focused on the problem than IPS, a DoS defense system (DDS) can block connection-based DoS attacks and those with legitimate content but bad intent. A DDS can also address both protocol attacks (such as teardrop and ping of death) and rate- based attacks (such as ICMP floods and SYN floods).

Blackholing and Sinkholing

With blackholing, all the traffic to the attacked DNS or IP address is sent to a "black hole" (null interface or a nonexistent server). To be more efficient and avoid affecting network connectivity, it can be managed by the ISP.

Sinkholing routes traffic to a valid IP address, which analyzes traffic and rejects bad packets. Sinkholing is not efficient for most severe attacks.

Upstream Filtering

All traffic is passed through a "cleaning center" or a "scrubbing center" via various methods such as proxies, tunnels, or even direct circuits, which separates "bad" traffic (DDoS and also other common Internet attacks) and only sends good traffic beyond to the server. The provider needs central connectivity to the Internet to manage this kind of service unless they happen to be located within the same facility as the "cleaning center" or "scrubbing center."

Examples of providers of this service include the following:

- Radware A10 Networks Arbor Networks AT&T F5 Networks Incapsula Neustar Inc. Prolexic Technologies
- Sprint Staminus Communications Tata Communications
- Verisign Verizon

Unintentional Denial of Service

An unintentional denial of service can occur when a system ends up denied, not due to a deliberate attack by a single individual or group of individuals, but simply due to a sudden enormous spike in popularity. This can happen when an extremely popular website posts a prominent link to a second, less prepared site, for example, as part of a news story. The result is that a significant proportion of the primary site's regular users—potentially hundreds of thousands of people—click that link in the space of a few hours, having the same effect on the target website as a DDoS attack. A VIPDoS is the same, but specifically when the link is posted by a celebrity.

When Michael Jackson died in 2009, websites such as Google and Twitter slowed down or even crashed. Many sites' servers thought the requests were from a virus or spyware trying to cause a denial-of-service attack, warning users that their queries looked like automated requests from a computer virus or spyware application.

News sites and link sites—sites whose primary function is to provide links to interesting content elsewhere on the Internet—are most likely to cause this phenomenon. The canonical example is the Slashdot effect when receiving traffic from Slashdot. It is also known as "the Reddit hug of death" and "the Digg effect."

Routers have also been known to create unintentional DoS attacks, as both D-Link and Netgear routers have overloaded NTP servers by flooding NTP servers without respecting the restrictions of client types or geographical limitations.

Similar unintentional denials of service can also occur via other media (e.g., when a URL is mentioned on television). If a server is being indexed by Google or another search engine during peak periods of activity, or does not have a lot of available bandwidth while being indexed, it can also experience the effects of a DoS attack.

Legal action has been taken in at least one such case. In 2006, Universal Tube & Rollform Equipment Corporation sued YouTube: Massive numbers of would-be YouTube users accidentally typed the tube company's URL, utube.com. As a result, the tube company ended up having to spend large amounts of money on upgrading their bandwidth. The company appears to have taken advantage of the situation, with utube.com now containing ads for advertisement revenue.

In March 2014, after Malaysia Airlines Flight 370 went missing, Digital-Globe launched a crowdsourcing service on which users could help search for the missing jet in satellite images. The response overwhelmed the company's servers.

An unintentional denial of service may also result from a prescheduled event created by the website itself. This could be caused when a server provides some service at a specific time. This might be a university website setting the grades to be available where it will result in many more login requests at that time than any other.

Side Effects of Attacks

Cyberattacks often result in additional security issues even after the main problem is resolved.

Backscatter

In computer network security, backscatter is a side effect of a spoofed denial-of-service attack. In this kind of attack, the attacker spoofs (or forges) the source address in IP packets sent to the victim. In general, the victim machine cannot distinguish between the spoofed packets and legitimate packets, so the victim responds to the spoofed packets as it normally would. These response packets are known as backscatter.

If the attacker is spoofing source addresses randomly, the backscatter response packets from the victim will be sent back to random destinations. This effect can be used by network telescopes as indirect evidence of such attacks.

The term "backscatter analysis" refers to observing backscatter packets arriving at a statistically significant portion of the IP address space to determine characteristics of DoS attacks and victims.

Legality

Many jurisdictions have laws under which denial-of-service attacks are illegal.

- In the US, denial-of-service attacks may be considered a federal crime under the Computer Fraud and Abuse Act, with penalties that include years of imprisonment. The Computer Crime and Intellectual Property Section of the US Department of Justice handles cases of (D)DoS.

- In European countries, committing criminal denial-of-service attacks may, at a minimum, lead to arrest. The United Kingdom is unusual in that it specifically outlawed denial-of-service attacks and set a maximum penalty of 10 years in prison with the Police and Justice Act of 2006, which amended Section 3 of the Computer Misuse Act of 1990.

On January 7, 2013, Anonymous posted a petition on the whitehouse.gov site asking that DDoS be recognized as a legal form of protest similar to the Occupy protests, the claim being the similarity in purpose.

2.7 SECURITY THREATS

2.7.1 Malicious Software

A computer program is a sequence of instructions to achieve a desired functionality. It is termed as malicious if the sequence of instructions is used to intentionally create adverse effects on the system. This type of code is malicious code, which is intended to harm, disrupt, or circumvent the computer system and the network resources.

Malicious code is new type of threat in form of auto-executable applications. It could be in form of scripting languages like Java Applets, ActiveX controls, and Java Script. The classification of such a code is given as follows:

- **Code that cause access violation:** Code that tries to delete, steal, alter, or execute unauthorized files. It can steal passwords, files, and other confidential data.

- **Code that enables DoS attacks:** Code that prevents the user from using the system. It may destroy the files that are open at the time of the attack.

2.7.2 Viruses

Computer viruses are the programs that can replicate their structures or effects by infecting other files or structures on a computer. The common use of a virus is to take over a computer to steal data.

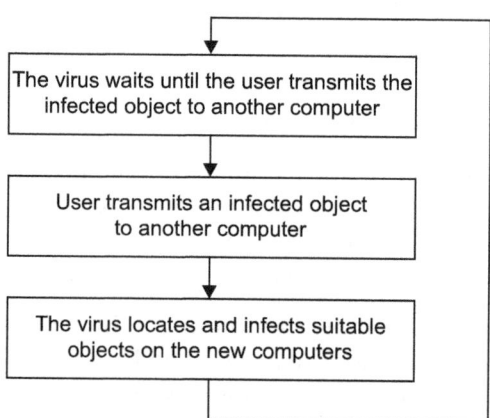

A Typical Life Cycle of a Computer Virus

Viruses come in many forms, thus breaking them into categories is not easy as many of them have multiple characteristics and may fall into multiple categories. We are going to divide them on basis of *what* they infect and *how* they infect.

Virus Categories on the Basis of What They Infect

S.No.	Category	Description
1.	System sector viruses	Viruses that infect control information on the disk itself.
2.	File viruses	Viruses that infect program .com and .exe files.
3.	Macro viruses	Viruses that infect data files that contains the macro programs like spell check in Microsoft Word files.
4.	Companion viruses	A special type of virus that adds files that run first to a disk.
5.	Cluster viruses	A special type of virus that infects through disk directory.
6.	Batch file viruses	Viruses that use text batch files to infect.
7.	Source code viruses	Viruses that add code to actual program code files.
8.	Visual Basic worms	Worms that use visual basic language code to control computer systems and perform tasks.

Virus Categories on the Basis of How They Infect

S.No.	Category	Description
1.	Polymorphic viruses	Viruses that change their characteristics as they infect.
2.	Stealth viruses	They try to actively hide themselves from anti-virus and system software.
3.	Fast and slow infectors	Viruses that change the activity to hide from anti-virus software.
4.	Sparse infectors	Viruses that do not infect very often.
5.	Armored infectors	Viruses that are programmed to make disassembly difficult.

(continued)

(continued)

S.No.	Category	Description
6.	Multipartite viruses	Viruses that fall into multiple top classes.
7.	Cavity (space filler) Viruses	Viruses that attempt to maintain a constant file size while infecting.
8.	Camouflage viruses	Viruses that attempt to appear as benign program to scenners.
9.	Tunneling viruses	Viruses that attempt to "tunnel" under anti-virus software while infecting.

2.7.3 Email Viruses

An email virus is computer code sent via an email note attachment, which, if activated, will cause some unexpected and usually harmful effect, such as destroying certain files on a hard disk and causing the attachment to be re-mailed to everyone in an address book. Although not the only kind of computer virus, email viruses are the best known and undoubtedly cause the greatest loss of time and money overall. The best two defenses against email viruses for individual users are as follows:

1. A policy of never opening (double clicking) an email attachment unless you know who sent it and what the attachment contains

2. Installing and using anti-virus software to scan any attachment before you open it

2.7.4 Macro Viruses

Macro viruses are a type of computer virus that are encoded as a macro embedded in a document. Many applications such as Microsoft Word and Excel support powerful macro languages. These applications allow us to embed a macro in a document and have the macro execute each time the document is opened.

According to some estimates, 75 percent of all viruses today are macro viruses. Once a macro virus gets into our machine, it can embed itself in all future documents we create with the application.

2.7.5 Worms

Computer worms are programs that can replicate themselves throughout a computer network, performing malicious tasks throughout.

A pure worm is more independent than a virus. A pure worm works by itself as an independent object. It does not need a carrier object to attach itself to. The worm can also spread by initiating telecommunications by itself. There is no need to wait for a human to send the file or document.

Mail Worms

A mail worm is carried by an email message, usually as an attachment, but there have been some cases when the worm is located in the message body. The recipient must open or execute the attachment before the worm can activate. The attachment may be a document with the worm attached in a virus-like manner, or it may be an independent file.

Pure Worms

Pure worms have the potential to spread very quickly because they are not dependent on any human actions, but the current networking environment is not ideal for them. They usually require a direct real-time connection between the source and target computer when the worm replicates.

2.7.6 Trojan Horse

A Trojan horse, commonly known as a Trojan, is a general term for malicious software that pretends to be harmless so that a user willingly allows it to be downloaded onto the computer. The name Trojan horse is borrowed from Greek mythology. In the computer world the term refers to a program that contains hidden malicious functions.

The program may look like something funny or useful such as a game or utility, but harms the system when executed. Trojans lack a replication routine and thus are not viruses by definition. A Trojan is spread to other computers only through deliberate transfer by users.

Backdoor Trojans

Backdoor Trojans are a special kind of Trojan that grant unauthorized access to computer systems. This type of Trojan is rather common and can pose a significant threat to business users. These Trojans consists of two programs that interoperate:

1. The silent server module planted in a victim's computer. The silent server module acts as a spying tool. The server module of a backdoor Trojan is often hidden in a useful program such as a game or utility.

2. The console based by a hacker.

2.7.7 Logic Bombs

Logic bombs refer to the program code that executes when a predefined event occurs. These logic bombs display a message to user and occur when a user is either accessing the Internet or making use of word-processing applications.

Logic bombs do not directly attack; however, they are responsible for informing victims if the criteria for an attack to start have been met.

2.7.8 Trapdoors

Trapdoors, also referred to as backdoors, are bits of code embedded in programs by programmers to quickly gain access at a later time, often during the testing or debugging phase. If an unscrupulous programmer purposely leaves this code in or simply forgets to remove it, a potential security hole is introduced. Hackers often plant a backdoor on previously compromised systems to gain later access. Trapdoors can be almost impossible to remove in a reliable manner. Often, reformatting the system is the only sure way.

2.7.9 Spoofs

Spoofing is the action of making something look like something that it is not in order to gain unauthorized access to a user's private information. The idea of spoofing originated in the 1980s with the discovery of a security hole in the TCP protocol. Today spoofing exists in various forms:

- IP spoofing
- URL or content spoofing
- Email spoofing
- Caller ID spoofing

IP Spoofing

Internet protocol (IP) is the protocol used for transmitting messages over the Internet. It is a network protocol operating at layer 3 of OSI model. IP spoofing is the act of manipulating the headers in a transmitted message to mask a hacker's true identity so that the message can appear as though it is from a trusted source.

The hacker manipulates the packet by using tools to modify the "source address" field. The source address is the IP address of the sender of the message; therefore, once an intruder forges this address and the destination server opens up a connection, numerous attacks can take place.

IP spoofing can be prevented by monitoring packets using network monitoring software. A filtering router could also be installed on the router, an ACL (access control list) is needed to block private addresses on the downstream interface. On the upstream interface, source addresses originating outside of the IP valid range will be blocked from sending spoofed information.

URL or Content Spoofing

URL (uniform resource locator) spoofing occurs when one website appears as if it is another. The URL that is displayed is the not the real URL of the site; therefore, the information is sent to a hidden Web address. Security patches are released by Web browsers, which add the feature of revealing the "true" URL of a site in the Web browser.

Email Spoofing

Email spoofing is the act of altering the header of an email so that the email appears to be sent from someone else.

Email spoofing can cause the following:

- Confusion or a person being discredited
- Social engineering (phishing)
- Hidden sender identity (spamming)

Caller ID Spoofing

Caller ID spoofing refers to the practice of causing the telephone network to display a number on recipient's caller ID display that is not that of the actual originating station.

2.8 SECURITY THREATS TO E-COMMERCE

E-commerce is basically exchanging and trading goods, more or less with services from corporations. It is bringing business sections to the ground of Internet.

Common Threats

- The base for the website is ultimately the servers and the computers. The threats against the security of online media are the Trojan horse, active content, and viruses.

- The most-prone area to attack the shopping software cart. The routines applicable here are online purchasing updates, tracking customers' details and accounts, and bunch of other services. It is all tied up in software.

- The next point where there is need for security is online payment transactions. The whole system works around accepting credit cards details through a gateway that is purely online.

- Other such threats are cracking, spoofing, eavesdropping, and root kills. The trends also show us threats like system unavailability and denial of service or power interruptions.

Common Vulnerabilities

- SQL injection

- Price manipulation

- Cross-site scripting

- Weak authentication and authorization

2.9 ELECTRONIC PAYMENT SYSTEMS

The emergence of e-commerce has created new financial needs that in many cases cannot be effectively fulfilled by traditional payment systems. Recognizing this, virtually all interested parties are exploring various types of electronic payment systems and issues surrounding electronic payment systems and digital currency. Broadly electronic payment systems can be classified into four categories:

- Online credit card payment system

- Online electronic cash system

- Electronic check system

- Smart cards–based electronic payment system

Each payment system has its advantages and disadvantages for customers and merchants. These payment systems have a number of requirements:

- Security

- Acceptability

- Convenience

- Cost

- Anonymity

- Control

- Traceability

Therefore, instead of focusing on technological specifications of various electronic payment systems, the researchers have distinguished electronic payment systems based on what is being transmitted over the network and have analyze the differences of each payment system by evaluating their requirements and characteristics and assessing the applicability of each system.

As payment is an integral part of the mercantile process, an electronic payment system is an integral part of e-commerce. The emergence of e-commerce has created new financial needs that in many cases cannot be effectively fulfilled by traditional payment systems. For instance, new types of peer-to-peer payment methods that allow individuals to e-mail payments to the other individual. Recognizing this, virtually all interested parties (e.g., academicians, government, business community and financial service providers) are exploring various types of electronic payment systems and issues surrounding electronic payment systems and digital currency. Some proposed electronic payment systems are simply electronic versions of existing payment systems such as checks and credit cards while others are based on digital currency technology and have the potential for definitive impact on today's financial and monetary system. Popular developers of electronic payment systems predict fundamental changes in the financial sector because of innovations in electronic payment systems. Therefore, electronic payment systems, and in particular methods of payment being developed to support electronic commerce, cannot be studied in isolation. Failure to place these developments into the proper context is likely to result in undue focus on the various experimental initiatives to develop electronic forms of payment without a proper reflection on the broader implications for the existing payment system.

2.9.1 The Concept of Electronic Payment

Payment systems that use electronic distribution networks have been a frequent practice in the banking and business sector since 1960s, especially for the transfer of large amounts of money. In the four decades that have passed since their appearance, important technological developments have taken place, which on the one hand have expanded the possibilities of electronic payment systems and on the other hand have created new business and social practice, which makes the use of these systems necessary. These changes, naturally, have affected the definition of electronic payment, which is evolving based on the needs of each period. In its most general form, the term "electronic payment" includes any payment to businesses, banks, or public services from citizens or businesses, which are executed through a telecommunication or electronic network using modem technology. It is obvious that, based on this definition, the electronic payments are the payments that are executed by the payer himself, whether the latter is a consumer or business, without the intervention of another natural person. Furthermore, the payment is made from a distance, without the physical presence of the payer and naturally it does include cash. By providing such a definition of electronic payment systems, researchers include the transfer of information concerning the parties' accounts involved in the e-commerce transactions, as well as the technological means of distribution channels through which the transaction is executed.

Size of Electronic Payment

Electronic payment systems are conducted in different e-commerce categories such as business to business (B2B), business to consumer (B2C), consumer to business (C2B), and consumer to consumer (C2C), each of which has special characteristics that depend on the value of order. Danial classified electronic payment systems as follows:

- Micro payment (less than $10) that is mainly conducted in C2C and B2C e-commerce

- Consumer payment that has a value between $10 and $500 is conducted in mainly in B2C e-commerce

- Business payments that have a value of more than $500 are conducted mainly in B2B e-commerce. B2B transactions account for 95 percent of e-commerce transactions, while others account about 5 percent (Turban et al, 2004).

- P2P, which is related to the C2C category of transactions, is relatively small due to its stiff usability.

Further, Cavarretta and de Silva (1995) identify three classes of typical electronic transactions:

- Tiny value transactions: Below $1
- Medium value transactions: Between $1 and $1,000
- Large value transactions: Above $1,000

Systems that can support tiny value transactions have to trade off between conveniences of transactions (the major part of a cost in an extremely cheap transaction) versus the security or durability of transactions. On the other side of the amount range, large value transactions will require highly secure protocols whose implementations are costly: they may be online and/or carry traceability information. Finally, nearly all the systems can perform medium-value transactions.

2.9.2 Conventional versus Electronic Payment Systems

To get into the depth of the electronic payment process, it is better to understand the processing of conventional or traditional payment systems. A conventional process of payment and settlement involves a buyer-to-seller transfer of cash or payment information (e.g., check and credit cards). The actual settlement of payment takes place in the final processing network. A cash payment requires buyer withdrawals from his or her bank account, a transfer of cash to the seller, and the seller deposit of payment to his or her account. Non-cash payment mechanisms are settled by adjusting (i.e., crediting and debiting) the appropriate accounts between banks based on payment information conveyed via check or credit card.

2.9.3 The Process of Electronic Payment Systems

Electronic payment systems have been in operation since the 1960s and have been expanding rapidly as well as growing in complexity. After the development of conventional payment systems, EFT (electronic funds transfer) based payment systems came into existence. It was the first electronic-based system, which did not depend on a central processing intermediary. An electronic fund transfer is a financial application of EDI (electronic data interchange), which sends credit card numbers or electronic checks via secure private networks between banks and major corporations. To use EFT to clear payments and settle accounts, an online payment service needs to add the capabilities to process orders, accounts, and receipts. But a change came with the development of digital currency. The nature of digital currency or electronic money mirrors that of paper money as a means of payment. As such, digital currency payment systems have the same advantages as currency payment, namely anonymity and convenience. As in other electronic payment systems (e.g., EFT based and intermediary based), here too security during the transaction and storage is a concern, although from a different perspective; for digital currency systems, double spending, counterfeiting, and storage become critical issues, whereas eavesdropping and liability are important for the notational funds transfer.

Payment Instructions

A payment instruction is an instance of payment with the details necessary to perform payment actions. For example, if a payment method's name is Visa, the payment instructions consist of Visa, the payment amount, and the cardholder's name, billing address, account number, and expiration date.

A shopper adds a payment instruction during the checkout flow. When the payment instruction is added, the PIAdd command runs and persists payment information to the payment tables. A payment instruction is created in the EDPPAYINST and PPCPAYINST tables. The protocol data is stored in the PPCEXTDATA table.

Payment Methods

Payment methods are the means by which payments are made. Payment methods include credit card brands, electronic checks, gift certificates, and manufacturer's coupons. The shopper selects the payment method on the order summary page.

- Payment methods have the following properties:
- Payment method name
- Priority
- Ability to be edited (whether the amount can be changed by an administrator or system)
- Amount limits

The payment policy is specified in the POLICY and POLICYDESC database tables. The properties are defined in the PaymentMethodConfiguration.xml file.

A shopper can use more than one payment method to pay for an order. The priority determines which payment method should be used first (e.g., if a gift certificate has a high priority and a credit card has a medium priority, the gift certificate is used first).

By default the following payment methods are available:

Credit Cards

- AMEX
- Mastercard
- VISA

Bill Me Later

- Offline transaction; customer is billed

Check

- Electronic check or automated clearing house (ACH) transaction

COD

- Cash on delivery

LOC

- Line of credit

2.9.4 Types of Electronic Payment Systems

With the growing complexities in e-commerce transactions, different electronic payment systems have appeared in last few years. At least dozens of electronic payment systems proposed or already in practice are found (Murthy 2002). The grouping can be made on the basis of what information is transferred online. Murthy explained six types of electronic payment systems:

1. PC banking

2. Credit cards

3. Electronic checks (I-checks)

4. Micro payments

5. Smart cards

6. E-cash

Kalakota and Whinston (1996) identified three types of electronic payment systems:

1. Digital token–based electronic payment systems

2. Smart card–based electronic payment systems

3. Credit-based electronic payment systems

Dennis (2001) put electronic payment systems into two categories:

1. Electronic cash

2. Electronic debit and credit card systems

Anderson (1998) broadly divided electronic payment systems into four categories:

1. Online credit card payment system
2. Electronic check system
3. Electronic cash system
4. Smart card–based electronic payment system

2.9.4.1 Online Credit Card Payment System

Credit cards are pieces of plastic whose holder has been granted a revolving credit line. This enables the holder to make purchases and/or cash advances up to pre-arranged limits. The credit granted can be settled in full by the end of a specific period or in part, with the balance taken as extended credit. Interest may be charged on the transaction amounts from the date of each transaction or only the external credit where credit granted has not been settled in full.

The card normally contains the cardholder's name and account number and other information encoded on the magnetic strip. Some credit cards may be used at ATMs. Credit cards began in the late 40s when banks began giving out paper certificates that could be used like cash in local stores. Frankline National Bank of New York used the first real credit card in 1951.

They seek to extend the functionality of existing credit cards for use as online shopping payment tools. This payment system has been widely accepted by consumers and merchants throughout the world and is by far the most popular method of payment, especially in the retail markets (Laudon and Traver, 2002). This form of a payment system has several advantages, which were not available through the traditional modes of payment. Some of the most important are the following:

- Privacy
- Integrity
- Compatibility
- Good transaction efficiency
- Acceptability
- Convenience
- Mobility
- Low financial risk
- Anonymity

Added to all these, to avoid the complexity associated with the digital cash or electronic checks, consumers and vendors are also looking at credit card payments on the Internet as a possible time-tested alternative. But, this payment system has raised several problems for consumers and merchants. Online credit card payments seek to address several limitations for merchants including lack of authentication, repudiation of charges, and credit card fraud. They also seek to address consumer fears about using credit cards such as having to reveal credit information at multiple sites and repeatedly having to communicate sensitive information over the Internet. The basic process of the online credit card payment system is very simple. If consumers want to purchase a product or service, they simply send their credit card details to the service provider involved, and the credit card organizations handle this payment like any other.

2.9.4.2 Electronic Check System

Electronic checks, also known as e-checks and I-checks, are used to make electronic payments between two parties through an intermediary and are not much different from the traditional or current check processing system. Electronic checks are generated and exchanged online. The intermediary will debit the customer account and credit the merchant account. Electronic checks are different from electronic funds transfer (EFT) is several ways. For electronic checking, electronic versions of checks are issued, received, and processed. So, the payee issues an electronic check for each payment. For EFT, automatic withdrawls are made for monthly bills or other fixed payments; no checks are issued. Thus, a complete electronic check transaction may consist of several basic steps, and these steps are executed in three distinct and optionally separate phases. In the first phase, the consumer makes a purchase; in the second phase, the merchants send the electronic checks to banks for redemption. In the third phase, the merchant's bank approaches the clearing house or consumer's bank to cash the electronic checks.

Electronic checks address the electronic needs of millions of businesses, which today exchange traditional paper checks with the other vendors, consumers, and government. The e-check method was deliberately created to work in much the same way as conventional paper checks. An account holder will issue an electronic document that contains the name of the financial institution, the payer's account number, the name of payee, and the amount of cheque. Most of the information is in uncoded form. Like a paper check e-checks also bear the digital equivalent of a signature and a computed number that authenticates the check from the owner of the account. Digital checking payment systems seek to extend the functionality of existing checking accounts for use as an online shopping payment tool. Electronic check systems have many advantages:

- They do not require consumers to reveal account information to other individuals when setting an auction.
- They do not require consumers to continually send sensitive information over the Web.
- They are less expensive than credit cards.
- They are much faster than paper-based traditional checks.

But this system also suffers from several disadvantages:

- Relatively high costs
- Limited used only in virtual world

The fact that they can prevent users' anonymity means they are not suitable for consumer retail transactions, although they are useful for government and B2B operations because the latter transactions do not require anonymity and the amount of transactions is generally large enough to cover fixed processing costs. The process of electronic checking can be described using the following steps:

1. A purchaser fills a purchase order form, attaches a payment advice (electronic check), signs it with his private key (using his signature hardware), attaches his public key certificate, encrypts it using his private key, and sends it to the vendor.

2. The vendor decrypts the information using his private key; checks the purchaser's certificates, signature, and cheque; attaches his deposit slip; and endorses the deposit by attaching his public key certificates. This is encrypted and sent to his bank.

3. The vendor's bank checks the signatures and certificates and sends the check for clearance. The banks and clearing houses normally have a private secure data network.

4. When the check is cleared, the amount is credited to the vendor's account and a credit advice is sent to him.

5. The purchaser gets a consolidated debit advice periodically.

2.9.4.3 Electronic Cash System

E–cash portability means that it must be freely transferable between any two parties in all forms of e-commerce transactions. In contrast, credit cards do not possess this property of portability or transferability between every combination of parties. In credit card transactions, the credit card payment recipient must already have a merchant's account established with a bank—a condition that is not required with electronic cash.

Electronic cash (e-cash) is a new concept in online payment systems because it combines computerized convenience with security and privacy that improves paper cash. Its versatility opens up a new host of markets and applications. E-cash is an electronic or digital form of value storage and value exchange that has limited convertibility into other forms of values that require intermediaries to convert. E-cash presents some characteristics like monetary value, storability and irretrievability, interoperability, and security. All these characteristics make it a more attractive payment system over the Internet. Added to these, this payment system offers numerous advantages like authority, privacy, good acceptability, low transaction cost, convenience, and good anonymity.

2.9.4.4 Smart Card–Based Electronic Payment Systems

Smart cards have received renewed attention as a mode of online payment. They are essentially credit card–sized plastic cards with money chips, and in some cases with micro processors embedded in them so as to serve as storage devices for much greater information than credit cards with built-in transaction processing capability.

These cards also contain an encrypted key that is compared to a secret key contained on the user's processor. Some smart cards have the provision to allow users to enter a personal identification number (PIN) code. Smart cards have been in use for well over two decades now and are widespread mostly in Europe and Asian countries. Owing to their considerable flexibility, they have been used for a wide range of functions like highway toll payments, prepaid phone cards, stored-value debit cards, and metro smart cards. However, with the recent emergence of e-commerce, these devices are increasingly being viewed as a particularly appropriate method to execute online payment systems with a considerably greater level of security than credit cards.

2.9.4.5 Factors of Electronic Payment Systems

- Integrity
- Non-repudiation

- Authentication
- Authorization
- Confidentiality
- Reliability

2.9.4.6 Comparison of Electronic Payment Systems (EPS)

Feature	Online credit card payment	Electronic cash	Electronic check	Smart cards
Actual payment time	Paid later	Prepaid	Paid later	Prepaid
Transaction information transfer	The store and bank checks the status of the credit card	Free transfer. No need to leave the name of involved	Electronic checks or payment indication must be endorsed	The smart card of both parties make the transfer
Online and offline transactions	Online transactions	Online transactions	Offline transfers are allowed	Offline transfers are allowed
Bank account involvement	Credit card account makes the payment	No involvement	The bank account makes the payment	The smart card account makes the payment
Users	Any legitimate credit card users	Anyone	Anyone with a bank account	Anyone with a bank or credit card account
Party to which payment is made out	Distributing Bank	Store	Store	Store
Consumer's transaction risk	The distributing bank bears the most risk; consumers only have to bear part of the risk	Consumer is at risk of the electronic cash getting stolen, lost, or misused	Consumers bear most of the risk, but the consumer can stop check payments at any time	Consumer is at risk of smart card getting stolen, lost, or misused

(continued)

Feature	Online credit card payment	Electronic cash	Electronic check	Smart cards
Current degree of popularity	Credit card organizations check for certification then total the purchases; can be used internationally and is the most popular payment type	Unable to meet financial Internet standards in the areas of expansion potential and internationalism	Cannot meet international standards and is therefore not very popular	Credit cards organizations check for certification then total the purchases; can be used internationally and is becoming more widely used
Anonymity	Partially or entirely anonymous	Entirely anonymous	No anonymity	Entirely anonymous, but if needed, the central processing agency can ask stores to provide information about a consumer
Small payments	Transactions costs are high; not suitable for small payments	Transactions costs are low; suitable for small payments	Allow stores to accumulate debt until it reaches a limit before paying for it; suitable for small payments	Transactions costs are low. Allows stores to communicate debt until it reaches a limit before paying for it. Therefore, it is suitable for small payments

(continued)

(continued)

Feature	Online credit card payment	Electronic cash	Electronic check	Smart cards
Database safeguarding	Safeguards regular credit card account	Needs to safeguard a large database and maintain records of the serial numbers of used electronic cash	Safeguards regular account information	Safeguards regular account information
Transaction information face value	Can be signed and issued freely in compliance with the limit	Face value is often set and cannot be changed	Can be signed and issued freely in compliance with the limit	Can be signed and issued freely in compliance with the limit
Real/Virtual world	Can be partially used in real world	Can only be used in virtual world	Limited to virtual world, but can share a checking account in the real world	Can be used in real or virtual worlds
Limit on transfer	Depends on the limit of credit card	Depends on how much is prepaid	No limit	Depends on how much money is saved
Mobility	Yes	No	No	Yes

2.10 DIGITAL SIGNATURE

The process of digitally signing starts by taking a mathematical summary (called a hash code) of the check. This hash code is a uniquely identifying digital fingerprint of the check. If even a single bit of the check changes, the hash code will dramatically change. The next step in creating a digital signature is to sign the hash code with a private key. This signed hash code is then appended to the check.

A digital signature is an electronic signature that can be used to authenticate the identity of sender of a message or the signer of a document, and possibly to ensure that the original content of the message or document that has been sent is unchanged. Digital signatures are easily transportable, cannot be imitated by someone else, and can be automatically stamped. The ability to ensure that the original signed message arrived means that the sender cannot easily repudiate it later.

A digital signature can be used with any kind of message, whether it is encrypted or not, simply so that the receiver can be sure of the sender's identity and that the message arrived intact. A digital certificate contains the digital signature of the certificate-issuing authority so that any one can verify that the certificate is real.

2.11 CRYPTOGRAPHY

Cryptography or *cryptology* is the practice and study of techniques for secure communication in the presence of third parties called adversaries. More generally, cryptography is about constructing and analyzing protocols that prevent third parties or the public from reading private messages; various aspects in information security such as data confidentiality, data integrity, authentication, and non-repudiation are central to modern cryptography. Modern cryptography exists at the intersection of the disciplines of mathematics, computer science, and electrical engineering. Applications of cryptography include ATM cards, computer passwords, and electronic commerce.

Cryptography prior to the modern age was effectively synonymous with encryption, the conversion of information from a readable state to apparent nonsense. The originator of an encrypted message (Alice) shared the decoding technique needed to recover the original information only with intended recipients (Bob), thereby precluding unwanted persons (Eve) from doing the same. The cryptography literature often uses Alice ("A") for the sender, Bob ("B") for the intended recipient, and Eve ("eavesdropper") for the adversary. Since the development of rotor cipher machines in World War I and the advent of computers in World War II, the methods used to carry out cryptology have become increasingly complex and its application is more widespread.

Modern cryptography is heavily based on mathematical theory and computer science practice; cryptographic algorithms are designed around computational hardness assumptions, making such algorithms hard to break in practice by any adversary. It is theoretically possible to break such a system, but it is infeasible to do so by any known practical means. These schemes are therefore termed "computationally secure"; theoretical advances (i.e., improvements in integer factorization algorithms, and faster computing technology) require these solutions to be continually adapted. There exist information-theoretically secure schemes that cannot be broken even with unlimited computing power—an example is the one-time pad—but these schemes are more difficult to implement than the best theoretically breakable but computationally secure mechanisms.

The growth of cryptographic technology has raised a number of legal issues in the information age. Cryptography's potential for use as a tool for espionage and sedition has led many governments to classify it as a weapon and to limit or even prohibit its use and export. In some jurisdictions where the use of cryptography is legal, laws permit investigators to compel the disclosure of encryption keys for documents relevant to an investigation. Cryptography also plays a major role in digital rights management and copyright infringement of digital media.

Until modern times, cryptography was referred to almost exclusively as encryption, which is the process of converting ordinary information (called plaintext) into unintelligible text (called cipher text). Decryption is the reverse, in other words, moving from the unintelligible cipher text back to plain text. A cipher (or cypher) is a pair of algorithms that create the encryption and the reversing decryption. The detailed operation of a cipher is controlled both by the algorithm and in each instance by a "key." The key is a secret (ideally known only to the communicants), usually a short string of characters, which is needed to decrypt the cipher text. Formally, a "cryptosystem" is the ordered list of elements of finite possible plain texts, finite possible cypher texts, finite possible keys, and the encryption and decryption algorithms that correspond to each key. Keys are important both formally and in actual practice, as ciphers without variable keys can be trivially broken with only the knowledge of the cipher used and are therefore useless (or even counter- productive) for most purposes. Historically, ciphers were often used directly for encryption or decryption without additional procedures such as authentication or integrity checks. There are two kinds of cryptosystems: symmetric and asymmetric. In symmetric systems the same key (the secret key) is used to encrypt and decrypt a message. Data manipulation in symmetric systems is faster than asymmetric systems as they generally use shorter key lengths. Asymmetric systems use a public key to encrypt a message and a private key to decrypt it. Use of asymmetric systems enhance the security of communication. Examples of asymmetric systems include RSA (Rivest-Shamir-Adleman), and ECC (elliptic curve cryptography). Symmetric models include the commonly used AES (advanced encryption system), which replaced the older DES (data encryption standard).

In colloquial use, the term "code" is often used to mean any method of encryption or concealment of meaning. However, in cryptography, code has a more specific meaning. It means the replacement of a unit of plaintext (i.e., a meaningful word or phrase) with a code word (for example, "wallaby" replaces "attack at dawn").

Cryptanalysis is the term used for the study of methods for obtaining the meaning of encrypted information without access to the key normally required to do so; it is the study of how to crack encryption algorithms or their implementations.

Some use the terms "cryptography" and "cryptology" interchangeably in English, while others (including US military practice, generally) use cryptography to refer specifically to the use and practice of cryptographic techniques and cryptology to refer to the combined study of cryptography and cryptanalysis. English is more flexible than several other languages in which cryptology (done by cryptologists) is always used in the second sense. RFC 2828 advises that steganography is sometimes included in cryptology.

The study of characteristics of languages that have some application in cryptography or cryptology (e.g., frequency data, letter combinations, universal patterns, etc.) is called cryptolinguistics.

History of Cryptography and Cryptanalysis

Before the modern era, cryptography was concerned solely with message of confidentiality (i.e., encryption)—conversion of messages from a comprehensible form into an incomprehensible one and back again at the other end, rendering it unreadable by interceptors or eavesdroppers without secret knowledge (namely the key needed for decryption of that message). Encryption attempted to ensure secrecy in communications, such as those of spies, military leaders, and diplomats. In recent decades, the field has expanded beyond confidentiality concerns to include techniques for message integrity checking, sender/receiver identity authentication, digital signatures, interactive proofs, and secure computation, among others.

Classic Cryptography

The earliest forms of secret writing required little more than writing implements since most people could not read. More literacy, or literate opponents, required actual cryptography. The main classical cipher types are transposition ciphers, which rearrange the order of letters in a message (e.g., "hello world" becomes "ehlol owrdl" in a trivially simple rearrangement scheme), and substitution ciphers, which systematically replace letters or groups of letters with other letters or groups of letters (e.g., "fly at once" becomes "gmz bu podf" by replacing each letter with the one following it in the Latin alphabet). Simple versions of either have never offered much confidentiality from enterprising opponents. An early substitution cipher was the Caesar cipher, in which each letter in the plain text was replaced by a letter some fixed number of positions further down the alphabet. Suetonius reports that Julius Caesar used it with a shift of three to communicate with his generals. Atbash is an example of an early Hebrew cipher. The earliest known use of cryptography is some carved cipher text on stone in Egypt, but this may have been done for the amusement of literate observers rather than as a way of concealing information.

The Greeks of classical times are said to have known of ciphers (e.g., the scytale transposition cipher claimed to have been used by the Spartan military). Steganography (i.e., hiding even the existence of a message so as to keep it confidential) was also first developed in ancient times. An early example, from Herodotus, was a message tattooed on a slave's shaved head and concealed under the regrown hair. More modern examples of steganography include the use of invisible ink, microdots, and digital watermarks to conceal information.

Cipher texts produced by a classical cipher (and some modern ciphers) will reveal statistical information about the plain text, and that information can often be used to break the cipher. After the discovery of frequency analysis, perhaps by the Arab mathematician and polymath Al-Kindi (also known as Alkindus) in the ninth century, nearly all such ciphers could be broken by an informed attacker. Such classical ciphers still enjoy popularity today, though mostly as puzzles (see cryptogram).

Language letter frequencies may offer little help for some extended historical encryption techniques such as homophonic ciphers that tend to flatten the frequency distribution. For those ciphers, language letter group (or n-gram) frequencies may provide an attack.

Essentially all ciphers remained vulnerable to cryptanalysis using the frequency analysis technique until the development of the polyalphabetic cipher, most clearly by Leon Battista Alberti around the year 1467, though there is some indication that it was already known to Al-Kindi. Alberti's innovation was to use different ciphers (i.e., substitution alphabets) for various parts of a message (perhaps for each successive plain-text letter at the limit). He also invented what was probably the first automatic cipher device, a wheel that implemented a partial realization of his invention. In the polyalphabetic Vigenère cipher, encryption uses a key word, which controls letter substitution depending on which letter of the key word is used. In the mid-nineteenth century Charles Babbage showed that the Vigenère cipher was vulnerable to Kasiski examination, but this was first published about ten years later by Friedrich Kasiski.

Although frequency analysis can be a powerful and general technique against many ciphers, encryption has still often been effective in practice, as many a would-be cryptanalyst was unaware of the technique. Breaking a message without using frequency analysis essentially required knowledge of the cipher used and perhaps of the key involved, thus making espionage, bribery, burglary, defection, and so on more attractive approaches to the cryptanalytically uninformed. It was finally explicitly recognized in the nineteenth century that secrecy of a cipher's algorithm is not a sensible nor practical safeguard of message security; in fact, it was further realized that any adequate cryptographic scheme (including ciphers) should remain secure even if the adversary fully understands the cipher algorithm itself. Security of the key used should alone be sufficient for a good cipher to maintain confidentiality under an attack. This fundamental principle was first explicitly stated in 1883 by Auguste Kerckhoffs and is generally called Kerckhoffs's principle; alternatively and more bluntly, it was restated by Claude Shannon, the inventor of information, the oryand, and the fundamentals of theoretical cryptography, as Shannon's Maxim—"The enemy knows the system."

Different physical devices and aids have been used to assist with ciphers. One of the earliest may have been the scytale of ancient Greece, a rod supposedly used by the Spartans as an aid for a transposition cipher. In medieval times, other aids were invented such as the cipher grille, which was also used for a kind of steganography. With the invention of polyalphabetic ciphers, came more sophisticated aids such as Alberti's own cipher disk, Johannes Trithemius's tabula recta scheme, and Thomas Jefferson's multi cylinder (not publicly known, and reinvented independently by Bazeries around 1900). Many mechanical encryption/decryption devices were invented early in the twentieth century, and several patented, among them rotor machines, famously including the Enigma machine used by the German government and military from the late 1920s and during World War II. The ciphers implemented by better-quality examples of these machine designs brought about a substantial increase in cryptanalytic difficulty after WWI.

Computer Era

Cryptanalysis of the new mechanical devices proved to be both difficult and laborious. In the United Kingdom, cryptanalytic efforts at Bletchley Park during WWII spurred the development of more efficient means for carrying out repetitious tasks. This culminated in the development of the Colossus, the world's first fully electronic, digital, programmable computer, which assisted in the decryption of ciphers generated by the German Army's Lorenz SZ40/42 machine.

Just as the development of digital computers and electronics helped in cryptanalysis, it made possible much more complex ciphers. Furthermore, computers allowed for the encryption of any kind of data representable in any binary format, unlike classical ciphers, which only encrypted written language texts; this was new and significant. Computer use has thus supplanted linguistic cryptography, both for cipher design and cryptanalysis. Many computer ciphers can be characterized by their operation on binary bit sequences (sometimes in groups or blocks), unlike classical and mechanical schemes, which generally manipulate traditional characters (i.e., letters and digits) directly. However, computers have also assisted cryptanalysis, which has compensated to some extent for increased cipher complexity. Nonetheless, good modern ciphers have stayed ahead of cryptanalysis; it is typically the case that use of a quality cipher is very efficient (i.e., fast and requiring few resources, such as memory or CPU capability), while breaking it requires an effort of many orders of larger magnitude, vastly larger than that required for any classical cipher, making cryptanalysis so inefficient and impractical as to be effectively impossible.

Extensive open academic research into cryptography is relatively recent; it began only in the mid-1970s. In recent times, IBM personnel designed the algorithm that became the federal (US) data encryption standard; Whitfield Diffie and Martin Hellman published their key agreement algorithm and the RSA algorithm in Martin Gardner's Scientific American column. Since then, cryptography has become a widely used tool in communications, computer networks, and computer security generally. Some modern cryptographic techniques can only keep their keys secret if certain mathematical problems are intractable, such as the integer factorization or the discrete logarithm problems, so there are deep connections with abstract mathematics. There are very few cryptosystems that are proven to be unconditionally secure. The one-time pad is one. There are a few important ones that are proven secure under certain unproven assumptions. For example, the infeasibility of factoring extremely large integers is the basis for believing that RSA is secure, but even there, the proof is usually lost due to practical considerations. There are systems similar to RSA such as one by Michael O. Rabin that is provably secure provided factoring n = pq is impossible, but the more practical system RSA has never been proved secure in this sense. The discrete logarithm problem is the basis for believing some other cryptosystems are secure, and again, there are related, less practical systems that are provably secure relative to the discrete log problem.

As well as being aware of cryptographic history, cryptograzphic algorithm and system designers must also sensibly consider probable future developments while working on their designs. For instance, continuous improvements in computer processing power have increased the scope of brute-force attacks, so when specifying key lengths, the required key lengths are similarly advancing. The potential effects of quantum computing are already being considered by some cryptographic system designers; the announced imminence of small implementations of these machines may be making the need for this preemptive caution rather more than being merely speculative.

Essentially, prior to the early twentieth century, cryptography was chiefly concerned with linguistic and lexicographic patterns. Since then the emphasis has shifted, and cryptography now makes extensive use of mathematics, including aspects of information theory, computational complexity, statistics, combinatorics, abstract algebra, number theory, and finite mathematics generally. Cryptography is also a branch of engineering, but an unusual one since it deals with active, intelligent, and malevolent opposition (see cryptographic engineering and security engineering); other kinds of engineering (e.g., civil or chemical engineering) need to deal only with neutral, natural forces. There is also active research examining the relationship between cryptographic problems and quantum physics (see quantum cryptography and quantum computer).

Symmetric-Key Cryptography

Symmetric-key cryptography refers to encryption methods in which both the sender and receiver share the same key (or, less commonly, in which their keys are different but related in an easily computable way). This was the only kind of encryption publicly known until June 1976.

Symmetric key ciphers are implemented as either block ciphers or stream ciphers. A block cipher enciphers input in blocks of plain text as opposed to individual characters, the input form used by a stream cipher.

The data encryption standard (DES) and the advanced encryption standard (AES) are block cipher designs that have been designated cryptography standards by the US government (though DES's designation was finally withdrawn after the AES was adopted). Despite its deprecation as an official standard, DES (especially its still-approved and much more secure triple-DES variant) remains quite popular; it is used across a wide range of applications, from ATM encryption to email privacy and secure remote access. Many other block ciphers have been designed and released, with considerable variation in quality. Many have been thoroughly broken, such as FEAL.

Stream ciphers, in contrast to the "block" type, create an arbitrarily long stream of key material, which is combined with the plaintext bit by bit or character by character, somewhat like the one-time pad. In a stream cipher, the output stream is created based on a hidden internal state, which changes as the cipher operates. That internal state is initially set up using the secret key material. RC4 is a widely used stream cipher. Block ciphers can be used as stream ciphers.

Cryptographic hash functions are a third type of cryptographic algorithm. They take a message of any length as input and output, a short, fixed-length hash that can be used in (for example) a digital signature. For good hash functions, an attacker cannot find two messages that produce the same hash. MD4 is a long-used hash function which is now broken; MD5, a strengthened variant of MD4, is also widely used but broken in practice. The US National Security Agency developed the Secure Hash Algorithm series of MD5-like hash functions: SHA-0 was a flawed algorithm that the agency withdrew; SHA-1 is widely deployed and more secure than MD5, but cryptanalysts have identified attacks against it; the SHA-2 family improves on SHA-1, but it isn't yet widely deployed; and the US standards authority thought it "prudent" from a security perspective to develop a new standard to "significantly improve the robustness of NIST's overall hash algorithm toolkit." Thus, a hash function design competition was meant to select a new US national standard, to be called SHA-3, by 2012. The competition ended on October 2, 2012, when the NIST announced that Keccak would be the new SHA-3 hash algorithm.

Message authentication codes (MACs) are much like cryptographic hash functions, except that a secret key can be used to authenticate the hash value upon receipt; this additional complication blocks an attack scheme against bare digest algorithms and so has been thought worth the effort.

Public-Key Cryptography

Symmetric-key cryptosystems use the same key for encryption and decryption of a message, though a message or group of messages may have a different key than others. A significant disadvantage of symmetric ciphers is the key management necessary to use them securely. Each distinct pair of communicating parties must, ideally, share a different key, and perhaps each cipher text exchanged as well. The number of keys required increases with the square of the number of network members, which very quickly requires complex key management schemes to keep them all consistent and secret. The difficulty of securely establishing a secret key between two communicating parties, when a secure channel does not already exist between them, also presents a chicken-and-egg problem, which is a considerable obstacle for cryptography users in the real world.

In a groundbreaking 1976 paper, Whitfield Diffie and Martin Hellman proposed the notion of public-key (also, more generally, called asymmetric key) cryptography in which two different but mathematically related keys are used—a public key and a private key. A public key system is so constructed that calculation of one key (the private key) is computationally infeasible from the other (the public key), even though they are necessarily related. Instead, both keys are generated secretly, as an interrelated pair. The historian David Kahn described public- key cryptography as "the most revolutionary new concept in the field since polyalphabetic substitution emerged in the Renaissance."

In public-key cryptosystems, the public key may be freely distributed, while its paired private key must remain secret. In a public-key encryption system, the public key is used for encryption, while the private or secret key is used for decryption. While Diffie and Hellman could not find such a system, they showed that public-key cryptography was indeed possible by presenting the Diffie–Hellman key exchange protocol, a solution that is now widely used in secure communications to allow two parties to secretly agree on a shared encryption key.

Diffie and Hellman's publication sparked widespread academic efforts in finding a practical public-key encryption system. This race was finally won in 1978 by Ronald Rivest, Adi Shamir, and Len Adleman, whose solution has since become known as the RSA algorithm.

The Diffie–Hellman and RSA algorithms, in addition to being the first publicly known examples of high-quality public-key algorithms, have been among the most widely used. Others include the Cramer–Shoup cryptosystem, ElGamal encryption, and various elliptic curve techniques.

To much surprise, a document published in 1997 by the Government Communications Headquarters (GCHQ), a British intelligence organization, revealed that cryptographers at GCHQ had anticipated several academic developments. Reportedly, around 1970, James H. Ellis had conceived the principles of asymmetric key cryptography. In 1973, Clifford Cocks invented a solution that essentially resembles the RSA algorithm. And in 1974, Malcolm J. Williamson is claimed to have developed the Diffie–Hellman key exchange.

Public-key cryptography can also be used for implementing digital signature schemes. A digital signature is reminiscent of an ordinary signature; they both have the characteristic of being easy for a user to produce, but difficult for anyone else to forge. Digital signatures can also be permanently tied to the content of the message being signed; they cannot then be "moved" from one document to another, for any attempt will be detectable. In digital signature schemes, there are two algorithms: one for signing, in which a secret key is used to process the message (or a hash of the message, or both), and one for verification, in which the matching public key is used with the message to check the validity of the signature. RSA and DSA are two of the most popular digital signature schemes. Digital signatures are central to the operation of public key infrastructures and many network security schemes (e.g., SSL/TLS, many VPNs, etc.).

Public-key algorithms are most often based on the computational complexity of "hard" problems, often from number theory. For example, the hardness of RSA is related to the integer factorization problem, while Diffie–Hellman key exchange protocol and DSA are related to the discrete logarithm problem. More recently elliptic curve cryptography has developed, a system in which security is based on number theoretic problems involving elliptic curves. Because of the difficulty of the underlying problems, most public-key algorithms involve operations such as modular multiplication and exponentiation, which are much more computationally expensive than the techniques used in most block ciphers, especially with typical key sizes. As a result, public-key cryptosystems are commonly hybrid cryptosystems, in which a fast, high-quality symmetric-key encryption algorithm is used for the message itself, while the relevant symmetric key is sent with the message, but encrypted using a public-key algorithm. Similarly, hybrid signature schemes are often used, in which a cryptographic hash function is computed and only the resulting hash is digitally signed.

Cryptanalysis

The goal of cryptanalysis is to find some weakness or insecurity in a cryptographic scheme, thus permitting its subversion or evasion.

It is a common misconception that every encryption method can be broken. In connection with his WWII work at Bell Labs, Claude Shannon proved that the one-time pad cipher is unbreakable, provided the key material is truly random, never reused, kept secret from all possible attackers, and of equal or greater length than the message. Most ciphers, apart from the one-time pad, can be broken with enough computational effort by brute force attack, but the amount of effort needed may be exponentially dependent on the key size, as compared to the effort needed to make use of the cipher. In such cases, effective security could be achieved if it is proven that the effort required ("work factor," in Shannon's terms) is beyond the ability of any adversary. This means it must be shown that no efficient method (as opposed to the time-consuming brute force method) can be found to break the cipher. Since no such proof has been found to date, the one-time-pad remains the only theoretically unbreakable cipher.

There are a wide variety of cryptanalytic attacks, and they can be classified in any of several ways. A common distinction turns on what Eve (an attacker) knows and what capabilities are available. In a cipher-text-only attack, Eve has access only to the cipher text (good modern cryptosystems are usually effectively immune to cipher-text-only attacks). In a known plain-text attack, Eve has access to a cipher text and its corresponding plain text (or to many such pairs). In a chosen plain-text attack, Eve may choose a plain text and learn its corresponding cipher text (perhaps many times); an example is gardening, used by the British during WWII. In a chosen cipher text attack, Eve may be able to choose cipher texts and learn their corresponding plain texts. Finally, in a man-in-the-middle attack, Eve gets in between Alice (the sender) and Bob (the recipient), accesses and modifies the traffic, and then forwards it to the recipient.

Cryptanalysis of symmetric key ciphers typically involves looking for attacks against the block ciphers or stream ciphers that are more efficient than any attack that could be against a perfect cipher. For example, a simple brute force attack against DES requires one known plain text and 2^{55} decryptions, trying approximately half of the possible keys, to reach a point at which chances are better than even that the key sought will have been found. But this may not be enough assurance; a linear cryptanalysis attack against DES requires 2^{43} known plain texts and approximately 2^{43} DES operations. This is a considerable improvement on brute force attacks.

Public-key algorithms are based on the computational difficulty of various problems. The most famous of these is integer factorization (e.g., the RSA algorithm is based on a problem related to integer factoring), but the discrete logarithm problem is also important. Much public-key cryptanalysis concerns numerical algorithms for solving these computational problems, or some of them, efficiently (i.e., in a practical time). For instance, the best-known algorithms for solving the elliptic curve-based version of discrete logarithms are much more time consuming than the best known algorithms for factoring, at least for problems of more or less equivalent size. Thus, other things being equal, to achieve an equivalent strength of attack resistance, factoring-based encryption techniques must use larger keys than elliptic curve techniques. For this reason, public-key cryptosystems based on elliptic curves have become popular since their invention in the mid-1990s.

While pure cryptanalysis uses weaknesses in the algorithms themselves, other attacks on cryptosystems are based on actual use of the algorithms in real devices and are called side-channel attacks. If a cryptanalyst has access to, for example, the amount of time the device took to encrypt a number of plain texts or report an error in a password or PIN character, he may be able to use a timing attack to break a cipher that is otherwise resistant to analysis. An attacker might also study the pattern and length of messages to derive valuable information; this is known as traffic analysis and can be quite useful to an alert adversary. Poor administration of a cryptosystem, such as permitting too-short keys, will make any system vulnerable, regardless of other virtues. And, of course, social engineering and other attacks against the personnel who work with cryptosystems or the messages they handle (e.g., bribery, extortion, blackmail, espionage, torture, etc.) may be the most productive attacks of all.

Cryptographic Primitives

Much of the theoretical work in cryptography concerns cryptographic primitives—algorithms with basic cryptographic properties—and their relationship to other cryptographic problems. More complicated cryptographic tools are then built from these basic primitives. These primitives provide fundamental properties, which are used to develop more complex tools called cryptosystems or cryptographic protocols, which guarantee one or more high-level security properties. Note, however, that the distinction between cryptographic primitives and cryptosystems is quite arbitrary; for example, the RSA algorithm is sometimes considered a cryptosystem and sometimes a primitive. Typical examples of cryptographic primitives include pseudorandom functions and one-way functions.

Cryptosystems

One or more cryptographic primitives are often used to develop a more complex algorithm, called a cryptographic system, or cryptosystem. Cryptosystems (e.g., El-Gamal encryption) are designed to provide particular functionality (e.g., public key encryption) while guaranteeing certain security properties (e.g., chosen plain-text attack (CPA) security in the random oracle model). Cryptosystems use the properties of the underlying cryptographic primitives to support the system's security properties. Of course, as the distinction between primitives and cryptosystems is somewhat arbitrary, a sophisticated cryptosystem can be derived from a combination of several more primitive cryptosystems. In many cases, the cryptosystem's structure involves back-and-forth communication among two or more parties in space (e.g., between the sender of a secure message and its receiver) or across time (e.g., cryptographically protected backup data). Such cryptosystems are sometimes called cryptographic protocols.

Some widely known cryptosystems include RSA encryption, Schnorr signature, El-Gamal encryption, and PGP. More complex cryptosystems include electronic cash systems and sign-cryption systems. Some of the more "theoretical" cryptosystems include interactive proof systems, (like zero-knowledge proofs) and systems for secret sharing.

Until recently, most security properties of most cryptosystems were demonstrated using empirical techniques or using ad hoc reasoning. Recently there has been considerable effort to develop formal techniques for establishing the security of cryptosystems; this has been generally called provable security. The general idea of provable security is to give arguments about the computational difficulty needed to compromise some security aspect of the cryptosystem (i.e., to any adversary).

The study of how best to implement and integrate cryptography in software applications is itself a distinct field (see cryptographic engineering and security engineering).

Prohibitions

Cryptography has long been of interest to intelligence gathering and law enforcement agencies. Secret communications may be criminal or even treasonous. Because of its facilitation of privacy, and the diminution of privacy attendant on its prohibition, cryptography is also of considerable interest to civil rights supporters. Accordingly, there has been a history of controversial legal issues surrounding cryptography, especially since the advent of inexpensive computers has made widespread access to high-quality cryptography possible.

In some countries, even the domestic use of cryptography is, or has been, restricted. Until 1999, France significantly restricted the use of cryptography domestically, though it has since relaxed many of these rules. In China and Iran, a license is still required to use cryptography. Many countries have tight restrictions on the use of cryptography. Among the more restrictive are laws in Belarus, Kazakhstan, Mongolia, Pakistan, Singapore, Tunisia, and Vietnam.

In the United States, cryptography is legal for domestic use, but there has been much conflict over legal issues related to cryptography. One particularly important issue has been the export of cryptography and cryptographic software and hardware. Probably because of the importance of cryptanalysis in World War II and an expectation that cryptography would continue to be important for national security, many Western governments have, at some point, strictly regulated export of cryptography. After World War II, it was illegal in the US to sell or distribute encryption technology overseas; in fact, encryption was designated as auxiliary military equipment and put on the United States munitions list. Until the development of the personal computer, asymmetric key algorithms (i.e., public key techniques), and the Internet, this was not especially problematic. However, as the Internet grew and computers became more widely available, high-quality encryption techniques became well known around the globe.

Export Controls

In the 1990s, there were several challenges to US export regulation of cryptography. After the source code for Philip Zimmermann's Pretty Good Privacy (PGP) encryption program found its way onto the Internet in June 1991, a complaint by RSA Security (then called RSA Data Security, Inc.) resulted in a lengthy criminal investigation of Zimmermann by the US Customs Service and the FBI, though no charges were ever filed. Daniel J. Bernstein, then a graduate student at UC Berkeley, brought a lawsuit against the US government challenging some aspects of the restrictions based on free speech grounds. The 1995 case *Bernstein v. United States* ultimately resulted in a 1999 decision that printed source code for cryptographic algorithms and systems was protected as free speech by the United States Constitution.

In 1996, thirty-nine countries signed the Wassenaar Arrangement, an arms control treaty that deals with the export of arms and "dual-use" technologies such as cryptography. The treaty stipulated that the use of cryptography with short key lengths (56-bit for symmetric encryption, 512-bit for RSA) would no longer be export controlled. Cryptography exports from the US became less strictly regulated as a consequence of a major relaxation in 2000; there are no longer very many restrictions on key sizes in US-exported mass-market software. Since this relaxation in US export restrictions, and because most personal computers connected to the Internet include US-sourced Web browsers such as Firefox or Internet Explorer, almost every Internet user worldwide has potential access to quality cryptography via their browsers (e.g., via transport layer security). The Mozilla Thunderbird and Microsoft Outlook email client programs similarly can transmit and receive emails via TLS and can send and receive email encrypted with S/MIME. Many Internet users don't realize that their basic application software contains such extensive cryptosystems. These browsers and email programs are so ubiquitous that even governments whose intent is to regulate civilian use of cryptography generally don't find it practical to do much to control distribution or use of cryptography of this quality, so even when such laws are in force, actual enforcement is often effectively impossible.

NSA Involvement

Another contentious issue connected to cryptography in the United States is the influence of the National Security Agency on cipher development and policy. The NSA was involved with the design of DES during its development at IBM and its consideration by the National Bureau of Standards as a possible federal standard for cryptography. DES was designed to be resistant to differential cryptanalysis, a powerful and general cryptanalytic technique known to the NSA and IBM that became publicly known only when it was rediscovered in the late 1980s. According to Steven Levy, IBM discovered differential cryptanalysis but kept the technique secret at the NSA's request. The technique became publicly known only when Biham and Shamir rediscovered and announced it some years later. The entire affair illustrates the difficulty of determining what resources and knowledge an attacker might actually have.

Another instance of the NSA's involvement was the 1993 Clipper chip affair, an encryption microchip intended to be part of the Capstone cryptography-control initiative. Clipper was widely criticized by cryptographers for two reasons. The cipher algorithm (called Skipjack) was then classified (declassified in 1998, long after the Clipper initiative lapsed). The classified cipher caused concerns that the NSA had deliberately made the cipher weak in order to assist its intelligence efforts. The whole initiative was also criticized based on its violation of Kerckhoffs's principle, as the scheme included a special escrow key held by the government for use by law enforcement, for example in wiretaps.

Digital Rights Management

Cryptography is central to digital rights management (DRM), a group of techniques for technologically controlling use of copyrighted material, being widely implemented and deployed at the behest of some copyright holders. In 1998, US President Bill Clinton signed the Digital Millennium Copyright Act (DMCA), which criminalized all production, dissemination, and use of certain cryptanalytic techniques and technology (now known or later discovered), specifically those that could be used to circumvent DRM technological schemes. This had a noticeable impact on the cryptography research community since an argument can be made that any cryptanalytic research violated, or might violate, the DMCA. Similar statutes have since been enacted in several countries and regions, including the implementation in the EU Copyright Directive. Similar restrictions are called for by treaties signed by World Intellectual Property Organization member-states.

The United States Department of Justice and FBI have not enforced the DMCA as rigorously as had been feared by some, but the law nonetheless remains a controversial one. Niels Ferguson, a well-respected cryptography researcher, has publicly stated that he will not release some of his research into an Intel security design for fear of prosecution under the DMCA. Both Alan Cox (longtime number 2 in Linux kernel development) and Edward Felten (and some of his students at Princeton) have encountered problems related to the act. Dmitry Sklyarov was arrested during a visit to the US from Russia and was jailed for five months pending trial for alleged violations of the DMCA, arising from work he had done in Russia, where the work was legal. In 2007, the cryptographic keys responsible for Blu-ray and HD DVD content scrambling were discovered and released onto the Internet. In both cases, the MPAA sent out numerous DMCA takedown notices, and there was a massive Internet backlash triggered by the perceived impact of such notices on fair use and free speech.

Forced Disclosure of Encryption Keys

In the United Kingdom, the Regulation of Investigatory Powers Act gives UK police the powers to force suspects to decrypt files or hand over passwords that protect encryption keys. Failure to comply is an offense in its own right, punishable on conviction by a two-year jail sentence or up to five years in cases involving national security. Successful prosecutions have occurred under the act; the first, in 2009, resulted in a term of 13 months' imprisonment. Similar forced disclosure laws in Australia, Finland, France, and India compel individual suspects under investigation to hand over encryption keys or passwords during a criminal investigation.

In the United States, the federal criminal case of *United States v. Fricosu* addressed whether a search warrant can compel a person to reveal an encryption passphrase or password. The Electronic Frontier Foundation (EFF) argued that this is a violation of the protection from self-incrimination given by the Fifth Amendment. In 2012, the court ruled that under the All Writs

Act, the defendant is required to produce an unencrypted hard drive for the court.

In many jurisdictions, the legal status of forced disclosure remains unclear.

The 2016 FBI–Apple encryption dispute concerns the ability of courts in the United States to compel manufacturers' assistance in unlocking cell phones whose contents are cryptographically protected.

EXERCISES

1. What are the considerations for *data security* in regard to *data backup, archival*, and *disposal*?

2. Discuss the various types of *firewalls*. How can we use firewalls to secure our applications?

3. What are *security threats*? Discuss the security threats to e-commerce.

4. What do you know about *digital signature*? How can we use it to authenticate and secure our digital documents?

5. What is *public key encryption*? Discuss in detail.

6. What, according to you, are the main threats for e-commerce security?

7. Explain in detail the two common modes of paying money online (i.e., e-cash and e-payment).

8. What do you understand by access control to information systems? How is authentication different from identification? Also differentiate between authentication and authorization.

9. What are the primary mechanisms for controlling access to information systems?

10. What are the primary considerations for *data security*? Discuss.

11. What are the different technologies used for data security and application? Explain the vulnerabilities associated with *VPN*.

12. What is a digital signature? Why do we use it?

13. What are *cryptographic attacks*? Explain in detail.

14. Write short notes on
   (a) IDS   (b) Client/server model
   (c) Three-tier architecture   (d) Hackers and crackers
   (e) Digital certificates   (f) NIDS

15. Comment on the importance of *digital signatures*. Explain how they are used to authenticate and secure digital documents.

16. What are the limitations of e-commerce?

17. Diagrammatically show the workings of *biometric technology*.

18. Prove the hypothesis "VPN is a boon."

19. What is meant by "fingerprint identification system"?

20. "Security is a fundamental concept of every network design." Justify the statement.

21. State the benefits of e-cash and electronic payment systems.

22. Define the term *data disposal* and add a note on its security aspects.

23. What is a *firewall*? How does the firewall work to protect an application?

24. Explain the concept of *cryptography*. What are the concepts of *public keys* and *private keys* while considering security of documents?

25. "To protect the valuable information in any government as well as private organization, security plays an important role." Explain the various threats and the protective measures one take to ensure security.

26. What are the advantages of a *virtual private network*? How does the system work? What is the tunneling concept in this regard?

27. What is *plastic money*? What are its advantages and disadvantages?

28. Write short notes on *credit cards and debit cards*.

29. What is *SET*? What is its importance? What are the strength and weakness of SET? Explain the operation of SET.

30. What are email security policies?

31. What are the challenges in establishment of secure networks? Discuss.

32. What is electronic commerce? What are benefits of electronic commerce for business firms? What precautions must be taken to ensure confidentiality of the document sent over the network?

33. What is access control of an informational resource? Discuss the differences between user authentication and user identification.

34. Discuss various types of intrusions possible in network systems. What are the approaches used for detection of intrusions?

35. List broad activities involved in the settlement of a business transaction in e-commerce. List issues related to the safety of business transactions on Web.

CHAPTER 3

DEVELOPING SECURE INFORMATION SYSTEMS

3.1 SECURE INFORMATION SYSTEM DEVELOPMENT

Including security early in the information system development cycle (SDLC) will usually result in less expensive and more effective security than adding it to an operational system. Thus, a framework for incorporating security into all phases of the SDLC process, from initiation to disposal, is required. This helps to select and acquire cost-effective security controls by explaining how to include information system security requirements in appropriate phases of the SDLC.

A general SDLC is discussed that includes the following phases:

- Initiation
- Acquisition/development
- Implementation
- Operations/maintenance
- Disposition

Each of these five phases includes a minimum set of security steps need to effectively incorporate security into a system during its development. An organization will either use the general SDLC described in this document or will have developed a tailored SDLC that meets their specific needs.

To be most effective, information security must be integrated into the SDLC from its inception. First, a description of the key security roles and responsibilities that are needed in most information system developments is provided. Second, sufficient information about the SDLC is provided to allow a person who is unfamiliar with the SDLC process to understand the relationship between information security and the SDLC.

Many methods exist that can be used by an organization to effectively develop an information system. A traditional SDLC is called a linear sequential model. This model assumes that the system will be delivered near the end of its life cycle. Another SDLC method uses the prototyping model, which is often used to develop an understanding of system requirements without actually developing a final operational system. More complex systems require iterative development models. More complex models have been developed and successfully used to address the evolving complexity of advanced and sometimes large information system designs. Examples of these more complex models are the following:

- Spiral model
- Component assembly model
- Concurrent development model

A general SDLC is discussed that includes the following phases.

3.1.1 Initiation

3.1.1.1 Security Categorization

Security categorization defines three levels (low, moderate, or high) of the potential impact on organizations or individuals should there be a breach of security (a loss of confidentiality, integrity, and availability). Security categorization standards assist organizations in making the appropriate selection of security controls for their information systems.

3.1.1.2 Preliminary Risk Assessment

Preliminary risk assessment results in an initial description of the basic security needs of the system. A preliminary risk assessment should define the threat environment in which the system will operate.

3.1.2 Acquisition/Development

3.1.2.1 Risk Assessment

Risk assessment analysis identifies the protection requirements for the system through a formal risk assessment process. This analysis builds on the initial risk assessment performed during the initiation phase, but will be more in depth and specific.

3.1.2.2 Security Functional Requirement Analysis

Security functional requirement analysis requirements that may include the following components:

1. System security environment (i.e., enterprise information security policy and enterprise security architecture)

2. Security functional requirements

3.1.2.3 Security Assurance Requirement Analysis

Security assurance requirement analysis addresses the developmental activities required and assurance evidence needed to produce the desired level of confidence that the information security will work correctly and effectively. The analysis, based on legal and functional security requirements, will be used as the basis for determining how much and what kinds of assurances are required.

3.1.2.4 Cost of Considerations and Reporting

Cost of considerations and reporting determines how much of the development cost can be attributed to information security over the life cycle of the system. These costs include hardware, software, personnel, and training.

3.1.2.5 Security Planning

Security planning ensures that agreed-on security controls, planned or in place, are fully documented. The security plan also provides a complete characterization or description of the information system as well as attachments or references to key documents supporting the agency's information security program (e.g., configuration management plan, contingency plan, incident response plan, security awareness and training plan, rules of behavior, risk assessment, security test and evaluation results, system interconnection agreements, security authorizations/accreditations, and plan of action and milestones).

3.1.2.6 Security Control Development

Security control development ensures that security controls described in the respective security plans are designed, developed, and implemented. For information systems currently in operation, the security plans for those systems may call the development of additional security controls to supplement the controls already in place or the modification of selected controls that are deemed to be less than effective.

3.1.2.7 Development Security Test and Evaluation

The development security test and evaluation ensures that security controls developed for a new information system are working properly and are effective. Some types of security controls (primarily those controls of a non-technical nature) cannot be tested and evaluated until the information system is deployed; these controls are typically management and operational controls.

3.1.2.8 Other Planning Components

Other planning components ensure that all necessary components of the development process are considered when incorporating security into the life cycle. These components include selection of the appropriate contract type, participation by all necessary functional groups within an organization, participation by the certifier and accreditor, and development and execution of necessary contracting plans and processes.

3.1.3 Implementation

3.1.3.1 Inspection and Acceptance

Inspection and acceptance ensures that the organization validates and verifies that the functionality described in the specification is included in the deliverables.

3.1.3.2 Security Control Integration

Security control integration ensures that security controls are integrated at the operational site where the information system is to be deployed for operation. Security control settings and switches are enabled in accordance with vendor instructions and available security implementation guidance.

Security certification ensures that the controls are effectively implemented through established verification techniques and procedures and gives organizations' officials confidence that the appropriate safeguards and countermeasures are in place to protect the organization's information system. Security certification also uncovers and describes the known vulnerabilities in the information system.

3.1.3.3 Security Certification

Security certification ensures that the controls are effectively implemented through established verification techniques and procedures and gives organizations' officials confidence that the appropriate safeguards and countermeasures are in place to protect the organization's information system. Security certification also uncovers and describes the known vulnerabilities in the information system.

3.1.3.4 Security Accreditation

Security accreditation provides the necessary security authorization of an information system to process, store, or transmit information that is required. This authorization is granted by a senior organization official and is based on the verified effectiveness of security controls to some agreed-on level of assurance and an identified residual risk to agency assets or operations.

3.1.4 Operations and Maintenance

3.1.4.1 Configuration Management and Control

Configuration management and control ensures adequate consideration of the potential security impacts due to specific changes to an information system or its surrounding environment. Configuration management and configuration control procedures are critical to establish an initial baseline of hardware, software, and firmware components for the information system and subsequently controlling and maintaining an accurate inventory of any changes to the system.

3.1.4.2 Continuous Monitoring

Continuous monitoring ensures that controls continue to be effective in their application through periodic testing and evaluation. Security control monitoring (i.e., verifying the continued effectiveness of those controls over time) and reporting the security status of the information system to appropriate agency officials is an essential activity of a comprehensive information security program.

3.1.5 Disposition

3.1.5.1 Information Preservation

Information preservation ensures that information is retained, as necessary, to conform to current legal requirements and to accommodate future technology changes that may render the retrieval method obsolete.

3.1.5.2 Media Sanitization

Media sanitization ensures that data is deleted, erased, and written over as necessary.

3.1.5.3 Hardware and Software Disposal

Hardware and software disposal ensures that hardware and software is disposed of as directed by the information system security officer. After discussing these phases and the information security steps in detail, the guide provides specifications, tasks, and clauses that can be used in an RFP to acquire information security features, procedures, and assurances.

3.2   APPLICATION DEVELOPMENT SECURITY

Integrating security in to application development life cycle is not an all-or-nothing decision, but rather a process of negotiation within policy, risk, and development requirements. Engaging security teams—in house or outsourced—during the definition stage of application development determines the security areas necessary to satisfy policy and risk tolerance in the context of the organization.

3.2.1 Initial Review

The first step is the initial review, which will allow the security team to assess initial risks. The security team should work with the development team to gain an understanding of the following:

- The purpose of the application in the context of its users and its market
- Its technical environment in terms of application development and deployment
- Policy delivers (regulatory and risk)
- Processes and procedures
- Business continuity requirements for application availability

3.2.2 Definition Phase: Threat Modeling

Threat modeling is the practice of working with developers to identify critical areas of applications dealing with sensitive information. The model is used to map information flow and identify critical areas of the application's infrastructure that require added security attention.

Once the application is modeled and the critical areas and entry points are identified, security teams should work with the developers to create mitigation strategies for potential vulnerabilities. Threat modeling should be created early in development life cycle of every project to achieve a secure foundation while using resources efficiently. This process should be followed throughout the development process as the application evolves in complexity.

3.2.3 Design Phase: Design Review

Application design reviews are an important step in identifying potential security risks at the early development stage. It is important that this review is conducted by an independent and objective moderator who is separate from the development team. This process involves reviewing application documents and interviewing developers and application owners. This will help keep the business purpose of the application in the forefront for analysis and later recommendations.

Reviews are held at each stage of the development process. This includes the beginning of the design phase before code is written, the end of each software developmental phase throughout the life cycle, and, finally, before the application goes live.

The chief design consideration to implement security are the following:

- Validating inputs
- Exception handling
- Applying cryptography
- Using random numbers

3.2.4 Development Phase: Code Review

During this phase, the development and coding of the system takes place. As modules and phases are complete, and once unit testing for each is finished, security testing against units should be conducted throughout the development process. This includes testing units and reviewing code for best security practices. During this phase, the focus shifts to the hardware and network environment, ensuring that segments and trust relationships are appropriate, servers are hardened at the operating system level, and application software is configured and administered securely.

3.2.5 Deployment Phase: Risk Assessment

While security reviews have been conducted throughout the cycle, at this point, a risk assessment done prior to deployment is a step toward benchmarking the live application. Once risk has been benchmarked for the "go live" application, a strategy for mitigation of any risk can be put into place.

3.2.6 Risk Mitigation

Risk mitigation involves prioritizing, evaluating, and implementing the controls that the security team identifies as necessary to mitigate vulnerabilities discovered during the risk-assessment stage. The least costly approach to implement the most appropriate control to reduce the risk to the organization is advisable. For example, risk can be assumed or reduced to an acceptable level, risk can be avoided by removing the cause, and risk can be transferred by using other options that compensate, such as purchasing insurance. The security team should work closely with the appropriate teams in the decision-making process on the most suitable mitigation options for each identified risk.

3.2.7 Benchmark

The next step is to benchmark the resulting application against industry standards to deliver a security scorecard. This allows executives to determine whether the security integration efforts are in line with industry averages and where there are gaps to improve. Many phases can be benchmarked and will correspond to one or more of the security criteria relevant to the organization.

3.2.8 Maintenance Phase: Maintain

In order to maintain the strong security posture established, it's important to consider employing periodic security checks of all critical applications and controls. Securing an application is adequate for that moment in time, but new risks are introduced every day that could affect its security.

3.3 INFORMATION SECURITY GOVERNANCE AND RISK MANAGEMENT

Although governance and security programs are discussed in our industry, not many organizations or security professionals fully understand all that is involved with each and the relationship between these two concepts.

It is not enough to have some security policies and then just concentrate on securing your network. To integrate security within business processes, an organization needs to have a robust information security program that maps to its business drivers, legal and regulatory requirements, and threat profile. The following sections provide an introduction to what information security governance and a security program are and how to get them deployed within any environment.

3.3.1 Information Security Governance

Information security governance is similar in nature to corporate and IT governance because there is overlapping functionality and goals among the three. All three work within an organizational structure of a company and have the same goals of helping to ensure that the company will survive and thrive; they just each have different focuses.

Information security governance is all of the tools, personnel, and business processes that ensure that security is carried out to meet an organization's specific needs. It requires organizational structure, roles and responsibilities, performance measurement, defined tasks, and oversight mechanisms.

3.3.2 Risk Management

Managing risk is a complex, multifaceted activity that requires the involvement of the entire organization—from senior leaders/executives providing the strategic vision and top-level goals and objectives for the organization to mid-level leaders planning, executing, and managing projects; to individuals on the front lines operating the information systems supporting the organization's missions/business functions.

Risk management is a comprehensive process that requires organizations to:

- frame risk,
- assess risk,
- respond to risk, and
- monitor risk.

3.3.2.1 Frame Risk

The first component of risk management addresses how organizations frame risk or establish a risk context, that is, describe the environment in which risk-based decisions are made. The purpose of the risk-framing component is to produce a risk management strategy that addresses how organizations intend to assess the risk, respond to risk, and monitor the risk—making explicit and transparent the risk perceptions that organizations routinely use in making both investment and operational decisions. The risk frame establishes a foundation for managing risk and delineates the boundaries for risk-based decisions within organizations.

- **Risk assumptions:** Assumptions about the threats, vulnerabilities, consequences/impact, and likelihood of occurrence that affect how risk is assessed, responded to, and monitored over time
- **Risk constraints:** Constraints on the risk assessment, response, and monitoring alternatives under consideration
- **Risk tolerance:** Levels of risk, types of risk, and degrees of risk uncertainty that are acceptable
- **Priorities and trade-offs:** The relative importance of missions/business functions, trade-offs among different types of risk that organizations face, time frames in which organizations must address risk, and any factors of uncertainty that organizations consider in risk responses

3.3.2.2 Assess Risk

The second component of risk management addresses how organizations assess risk within the context of the organizational risk frame. The purpose of the risk assessment component is to identify the following:

- **Threats to organizations:** Operations, assets, or individuals or threats directed through organizations against other organizations or the nation
- **Vulnerabilities:** Internal and external with respect to an organization
- **The harm:** Consequences and impact to organizations that may occur given the potential for threats for exploiting vulnerabilities
- **The likelihood that harm will occur:** The end result is a determination of risk (i.e., the degree of harm and likelihood of harm occurring)

3.3.2.3 Respond to Risk

The third component of risk management addresses how organizations respond to risk once that risk is determined based on the results of risk assessments. The purpose of the risk response component is to provide a consistent, organization-wide response to risk in accordance with the organizational risk frame by the following:

- Developing alternative courses of action for responding to risk
- Evaluating the alternative courses of action
- Determining appropriate courses of action consistent with organizational risk tolerance
- Implementing risk responses based on selected courses of action

3.3.2.4 Monitor Risk

The fourth component of risk management addresses how organizations monitor risk over time. The purpose of the risk monitoring component is to do the following:

- Verify the planned risk response measures are implemented and information security requirements derived from/traceable to organizational mission business functions, federal legislation, directives, regulations, policies, and standards, and guidelines are satisfied
- Determine the ongoing effectiveness of risk response measures following implementation
- Identify risk-impact changes to organizational information systems and the environments in which the systems operate

The figure shows the risk management process. The bidirectional nature of the arrows indicates that the information and communication flows among the risk management components as well as the execution order of the components, and may be flexible and respond to the dynamic nature of the risk management process.

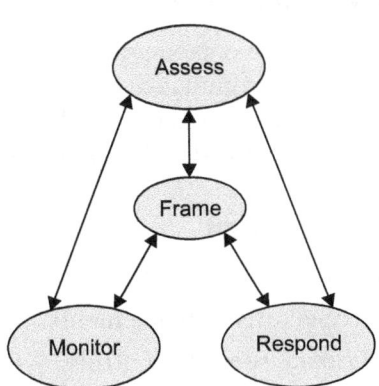

3.4 SECURITY ARCHITECTURE AND DESIGN

Security architecture is one component of a product's/system's overall architecture and is developed to provide guidance during the design of the product/system.

Security architecture is the design artifacts that describe how security controls (security countermeasures) are positioned and how they relate to the overall system's architecture. These controls serve the purpose to maintain the system's quality attributes such as confidentiality, integrity, and availability.

A security policy is a statement that outlines how entities access each other, what operations different entities can carry out, what level of protection is required for a system or software product, and what actions should be taken when these requirements are not met.

A security model outlines the requirements necessary to properly support and implement a certain security policy.

3.4.1 Computer System Architecture

A computer system consists of different types of components like hardware, software, operating systems, and firmware. The most important and common components are the following:

- Hardware components like CPU, storage devices, I/O devices, and communication devices
- Software components like operating systems and application programs
- Firmware

3.4.2 Systems Security Architecture

The security architecture is one component of a product's overall architecture and is developed to provide guidance during the design of the product. It outlines the level of assurance that is required and potential impacts that level of security could have during the development stages and on the product overall.

3.4.3 Principles of Secure Design

- Design security in from the start.
- Allow for future security enhancements.
- Minimize and isolate security controls.

- Employ least privilege.
- Structure the security relevant features.
- Make security friendly.
- Don't depend on secrecy for security.

3.4.4 Principles of Software Security

- Secure the weakest link.
- Practice defense in depth.
- Fail securely; if your software has to fail, make sure it does it securely.
- Follow the principle of least privilege.
- Compartmentalize/minimize the amount of damage that can be done by breaking the system into units.
- Keep it simple; complex design is never easy to understand.
- Promote privacy; try not to do anything that compromises the privacy of user.
- Remember that hiding secrets is hard.
- Be reluctant to trust; instead of making assumptions that need to hold true, you should be reluctant to extend trust.
- Use your community resources; public scrutiny promotes trust.

3.4.5 Security Product Evaluation Methods and Criteria

A security evaluation examines the security-relevant parts of a system, meaning the TCB, access control mechanisms, reference monitor, kernel, and protection mechanisms. The relationship and interaction between these components are also evaluated. There are different methods of evaluating and assigning assurance levels to systems.

Two reasons explain why more than one type of assurance evaluation process exist:

1. Methods and ideologies have evolved over time.

2. Various parts of the world look at computer security differently and rate some aspects of security differently.

An evaluation program establishes a trust between the security product vendor and the customer.

3.4.5.1 Evaluation Standard

The *Information Technology Security Evaluation Criteria (ITSEC)* is a structured set of criteria for evaluating computer security within products and systems. The ITSEC was first published in May 1990 in France, Germany, the Netherlands, and the United Kingdom based on existing work in their respective countries. Following extensive international review, Version 1.2 was subsequently published in June 1991 by the Commission of the European Communities for operational use within evaluation and certification schemes.

Since the launch of the ITSEC in 1990, a number of other European countries have agreed to recognize the validity of ITSEC evaluations.

The ITSEC has been largely replaced by Common Criteria, which provides similarly defined evaluation levels and implements the target of evaluation concept and the "Security Target" document.

Concept

The product or system being evaluated, called the target of evaluation, is subjected to a detailed examination of its security features, culminating in comprehensive and informed functional and penetration testing. The degree of examination depends on the level of confidence desired in the target. To provide different levels of confidence, the ITSEC defines evaluation levels, denoted E0 through E6. Higher evaluation levels involve more extensive examination and testing of the target.

Unlike earlier criteria, notably the TCSEC developed by the US defense establishment, the ITSEC did not require evaluated targets to contain specific technical features in order to achieve a particular assurance level. For example, an ITSEC target might provide authentication or integrity features without providing confidentiality or availability. A given target's security features were documented in a "Security Target" document, whose contents had to be evaluated and approved before the target itself was evaluated. Each ITSEC evaluation was based exclusively on verifying the security features identified in the "Security Target" document.

Use

The formal Z notation was used to prove security properties about the Mondex smart card electronic cash system, allowing it to achieve ITSEC level E6, the highest granted security-level classification.

The *Trusted Computer System Evaluation Criteria (TCSEC)* is a United States Government Department of Defense (DoD) standard that sets basic requirements for assessing the effectiveness of computer security controls built into a computer system. The TCSEC was used to evaluate, classify, and select computer systems being considered for the processing, storage, and retrieval of sensitive or classified information.

The TCSEC, frequently referred to as the *Orange Book*, is the centerpiece of the DoD Rainbow Series publications. Initially it was issued in 1983 by the National Computer Security Center (NCSC), an arm of the National Security Agency, and then it was updated in 1985. TCSEC was replaced by the Common Criteria international standard, originally published in 2005.

Fundamental Objectives and Requirements

The Orange Book or DoDD 5200.28-STD was canceled by DoDD 8500.1 on October 24, 2002. DoDD 8500.1 reissued as DoDI 8500.02 on March 14, 2014.

Policy

The security policy must be explicit, well defined, and enforced by the computer system. There are three basic security policies:

1. **Mandatory security policy:** Enforces access control rules based directly on an individual's clearance, authorization for the information, and the confidentiality level of the information being sought. Other indirect factors are physical and environmental. This policy must also accurately reflect the laws, general policies, and other relevant guidance from which the rules are derived.

2. **Marking:** Systems designed to enforce a mandatory security policy must store and preserve the integrity of access control labels and retain the labels if the object is exported.

3. **Discretionary security policy:** Enforces a consistent set of rules for controlling and limiting access based on identified individuals who have been determined to need to know the information.

Accountability

Individual accountability regardless of policy must be enforced. A secure means must exist to ensure the access of an authorized and competent agent who can then evaluate the accountability information within a reasonable amount of time and without undue difficulty. There are three requirements under the accountability objective:

1. **Identification:** The process used to recognize an individual user

2. **Authentication:** The verification of an individual user's authorization to specific categories of information

3. **Auditing:** Audit information selectively kept and protected so that actions affecting security can be traced to the authenticated individual

Assurance

The computer system must contain hardware/software mechanisms that can be independently evaluated to provide sufficient assurance that the system enforces the listed requirements. By extension, assurance must include a guarantee that the trusted portion of the system works only as intended. To accomplish these objectives, three types of assurance are needed with their respective elements:

Assurance Mechanisms

1. **Operational assurance:** System architecture, system integrity, covert channel analysis, trusted facility management, and trusted recovery

2. **Life-cycle assurance:** Security testing, design specification and verification, configuration management, and trusted system distribution

3. **Continuous protection assurance:** The trusted mechanisms that enforce these basic requirements must be continuously protected against tampering and/or unauthorized changes.

Documentation

Within each class there is additional documentation set, which addresses the development, deployment, and management of the system rather than its capabilities. This documentation includes the following:

1. Security features user's guide

2. Trusted facility manual

3. Test documentation

4. Design documentation

Divisions and Classes

The TCSEC defines four divisions: D, C, B, and A, where division A has the highest security. Each division represents a significant difference in the trust an individual or organization can place on the evaluated system. Additionally divisions C, B, and A are broken into a series of hierarchical subdivisions called classes: C1, C2, B1, B2, B3, and A1.

Each division and class expands or modifies as indicated by the requirements of the immediately prior division or class.

D: Minimal Protection

This level is reserved for those systems that have been evaluated but that fail to meet the requirements for a higher division.

C: Discretionary Protection

1. **C1: Discretionary security protection**
   a. Identification and authentication
   b. Separation of users and data
   c. Discretionary access control (DAC) capable of enforcing access limitations on an individual basis
   d. Required system documentation and user manuals

2. **C2: Controlled access protection**
   a. More finely grained DAC
   b. Individual accountability through log-in procedures
   c. Audit trails
   d. Object reuse
   e. Resource isolation
   f. An example is HP-UX

B: Mandatory Protection

1. **B1: Labeled security protection**
   a. Informal statement of the security policy model
   b. Data sensitivity labels
   c. Mandatory access control (MAC) over selected subjects and objects
   d. Label exportation capabilities

    e. Some discovered flaws must be removed or otherwise mitigated

    f. Design specifications and verification

2. **B2: Structured protection**

    a. Security policy model clearly defined and formally documented

    b. DAC and MAC enforcement extended to all subjects and objects

    c. Covert storage channels are analyzed for occurrence and bandwidth

    d. Carefully structured into protection-critical and non-protection-critical elements.

    e. Design and implementation enable more comprehensive testing and review

    f. Authentication mechanisms are strengthened

    g. Trusted facility management is provided with administrator and operator segregation

    h. Strict configuration management controls are imposed

    i. Operator and administrator roles are separated

    j. An example is Multics

3. **B3: Security domains**

    a. Satisfies reference monitor requirements

    b. Structured to exclude code not essential to security policy enforcement

    c. Significant system engineering directed toward minimizing complexity

    d. Security administrator role defined

    e. Security-relevant events audited

    f. Automated imminent intrusion detection, notification, and response

    g. Trusted system recovery procedures

    h. Covert timing channels analyzed for occurrence and bandwidth

    i. An example is the XTS-300, a precursor to the XTS-400

A: Verified Protection

1. **A1: Verified design**

   a. Functionally identical to B3

   b. Formal design and verification techniques including a formal top-level specification

   c. Formal management and distribution procedures

   d. Examples: Honeywell's SCOMP, Aesec's GEMSOS, and Boeing's SNS server. Two that were unevaluated were the production LOCK platform and the cancelled DEC VAX Security Kernel.

2. **Beyond A1**

   a. System architecture demonstrates that the requirements of self-protection and completeness for reference monitors have been implemented in the trusted computing base (TCB).

   b. Security testing automatically generates test case from the formal top-level specification or formal lower-level specifications

   c. Formal specification and verification is where the TCB is verified down to the source code level, using formal verification methods where feasible

   d. Trusted design environment is where the TCB is designed in a trusted facility with only trusted (cleared) personnel

Common Criteria: Hybrid of ITSEC and TCSEC

A security evaluation examines the security-relevant parts of a system, meaning the TCB, access control mechanisms, reference monitor, kernel, and protection mechanisms. The relationship and interaction between these components are also evaluated.

There are different methods of evaluating and assigning assurance levels to systems. Two reasons explain why more than one type of assurance evaluation process exist: Methods and ideologies have evolved over time, and various parts of the world look at computer security differently and rate some aspects of security differently.

An evaluation program establishes trust between the security product vendor and the customer.

The Criteria

1. **Security policy:** The policy must be explicit and well defined and enforced by the mechanisms within the system.

2. **Identification:** Individual subjects must be uniquely identified.

3. **Labels access:** Control labels must be associated properly with objects.

4. **Documentation:** Documentation must be provided, including test, design and specification documents, user guides, and manuals.

5. **Accountability:** Audit data must be captured and protected to enforce accountability.

6. **Life-cycle assurance:** Software, hardware, and firmware must be able to be tested individually to ensure that each enforces the security policy in an effective manner throughout their lifetimes.

7. **Continuous protection:** The security mechanisms and the system as a whole must perform predictably and acceptably in different situations continuously.

TCSEC Myths

a. It looks specifically at the operating system and not at other issues like networking and databases.

b. It focuses mainly on one attribute of security, confidentiality, and not at integrity and availability.

c. It works with government classifications and not the protection classifications that commercial industries use.

d. It has a relatively small number of ratings, which means many different aspects of security are not evaluated and rated independently.

TCSEC versus ITSEC

a. TCSEC bundles functionality and assurance into one rating, whereas ITSEC evaluates these two attributes separately.

b. ITSEC provides more flexibility than TCSEC.

c. ITSEC addresses integrity, availability, and confidentiality whereas TCSEC addresses only confidentiality.

d. ITSEC also addresses networked systems, whereas TCSEC deals with standalone systems.

Common Criteria

Overview

a.   Common Criteria was the result of ISO's attempt to gather several organizations to come together to combine and align existing and emerging evaluation criteria like TCSEC, ITSEC, Canadian Trusted Computer Product Evaluation Criteria (CTCPEC), and the federal criteria.

b.   It was developed through a collaboration among national security standards organizations within the United States, Canada, France, Germany, the United Kingdom, and the Netherlands.

c.   Under the Common Criteria model, an evaluation is carried out on a product and is assigned an evaluation assurance level (EAL).

Benefits of the Common Criteria

a.   Helps consumers by reducing the complexity of the ratings and eliminating the need to understand the definition and meaning of different ratings within various evaluation schemes.

b.   Helps manufacturers because now they can build to one specific set of requirements if they want to sell their products internationally, instead of having to meet several different ratings with varying rules and requirements.

CC Assurance Levels

The Common Criteria has seven assurance levels, which range from EAL1 (lowest), where functionality testing takes place, through EAL7 (highest), where thorough testing is performed and the system design is verified.

1.   EAL 1: Functionally tested

2.   EAL 2: Structurally tested

3.   EAL 3: Methodically tested and checked

4.   EAL 4: Methodically designed, tested, and reviewed

5.   EAL 5: Semi-formally designed and tested

6.   EAL 6: Semi-formally verified design and tested

7.   EAL 7: Formally verified design and tested

Protection Profile

A protection profile is a mechanism that is used by CC in its evaluation process to describe a real-world need of a product that is not currently on the market.

Protection Profile Characteristics

a. Contains the set of security requirements, their meaning and reasoning, and the corresponding EAL rating that the intended product will require.

b. Describes the environmental assumptions, the objectives, and the functional and assurance level expectations. Each relevant threat is listed along with how it is to be controlled by specific objectives.

c. Justifies the assurance level and requirements for the strength of each protection mechanism.

d. Provides a means for a consumer, or others, to identify specific security needs; this is the security problem that is to be conquered.

e. Provides the necessary goals and protection mechanisms to achieve the necessary level of security and a list of the things that can go wrong during the type of system development.

f. **Protection Profile Sections**

(i.) Descriptive elements: Provides the name of the profile and a description of the security problem that is to be solved.

(ii.) Rational: Justifies the profile and gives a more detailed description of the real-world problem to be solved. The environment, usage assumptions, and threats are illustrated along with guidance on the security policies that can be supported by products and systems that conform to this profile.

(iii.) Functional requirements: Establishes a protection boundary, meaning the threats or compromises that are within this boundary to be countered. The product or system must enforce the boundary established in this section.

(iv.) Development assurance: Requirements identify the specific requirements that the product or system must meet during the development phases, from design to implementation.

(v.) Evaluation assurance: Requirements establish the type and intensity of the evaluation.

CC Components

1. **Protection profile:** Description of needed security solution

2. **Target of evaluation:** Product proposed to provide needed security solution

3. **Security target:** Vendor's written explanation of the security functionality and assurance mechanisms that meet the needed security solution; in other words, "This is what our product does and how it does it."

4. **Packages—EALs:** Functional and assurance requirements are bundled into packages for reuse. This component describes what must be met to achieve specific EAL ratings.

Certification and Accreditation

Certification

a. Certification is the comprehensive technical evaluation of the security components and their compliance for the purpose of accreditation.

b. A certification process may use safeguard evaluation, risk analysis, verification, testing, and auditing techniques to assess the appropriateness of a specific system.

c. The goal of a certification process is to ensure that a system, product, or network is right for the customer's purposes.

Accreditation

a. Accreditation is the formal acceptance of the adequacy of a system's overall security and functionality by management.

b. Certification is a technical review that assesses the security mechanisms and evaluates their effectiveness. Accreditation is management's official acceptance of the information in the certification process findings.

3.5 SECURITY ISSUES IN HARDWARE, DATA STORAGE, AND DOWNLOADABLE DEVICES

3.5.1 Hardware

Unfortunately, theft of computer hardware from businesses is common. Although laptops and other portable devices are the easiest to steal, in some cases, thieves want internal devices such as processors and chips, so don't assume that less mobile equipment will be any safer.

- Site security
- Theft of equipment or intellectual property
- Loss or corruption of data through viruses
- Unauthorized entry into the computer system
- Accidental loss or corruption of data by a user
- Loss of private company or customer information

3.5.2 Security Marking

Security should mark all significant items of computer hardware, using special ultraviolet pens or embossed labels. Allocate serial numbers to all of your hardware and keep a record of these numbers.

3.5.3 Security Data Backup

While the theft of hardware is inconvenient, you can replace it. However, the loss of a PC or server also means you may have lost potentially critical business data. Therefore, it is important to take security measures for both hardware and data.

3.5.4 Power Supply Backup

An uninterruptible power supply makes sure that your key components, such as servers and network components, will continue to operate for a short time even if there is power is cut. This will give you time to shut systems down in an orderly fashion.

3.5.5 Data Storage

Data storage security is a wide-ranging area that covers everything from legal compliance to preparedness for e-discovery requests, to user access control and the physical security of data storage.

Physical threats to data centers cross a gamut of unpleasant possibilities. Most of them fall into one of three major classifications:

- Natural disasters
- Physical intrusions
- Energy issues

When we think about physical threats to the data center, our minds naturally go to the dramatic natural disasters: earthquakes, tornados, hurricanes, extreme weather, and tidal waves (Japan, Thailand).

Data centers should be located as far away from active disaster threats as possible. This does not mean you must build a data center across the country from your headquarters or contract with a data center provider many states away from your IT staff. For example, Raging Fire wanted to build close enough to its co-location customers in the Bay area although the region is seismically active. They built instead in less quake-prone Sacramento, inland of San Francisco and not too far for customer IT to visit.

3.6 PHYSICAL SECURITY OF IT ASSETS

An almost unlimited number of threats can theoretically be of concern to an organizations well-being. The main focus of physical security has often been on human-made disasters, such as attack over the network from a outside hacker or plain human error from employees. Even if these in fact are the most common threats, it is crucial not to forget that the same kind of threats also can occur from natural disasters such as fire, water, leakage, electricity disturbance, or other environmental failures.

3.6.1 The Human Factor

It is assumed among many computer security officials that a majority of theft, fraud, sabotage, and accidents are caused by a company's own employees. The most probable causes are accidental human error or tampering from unauthorized users.

3.6.2 Natural Disasters

As mentioned earlier, the main focus for physical access exposures is often given to the human factor. But it is very important to consider all potential threats, even unlikely ones. Natural disasters can actually have a major effect if they occur. Some of the natural disasters that can occur are the following:

- Fire
- Electrical interruption
- Lightning
- Environmental disaster

- Earthquake
- Water

3.6.3 Physical Access Control

Physical security controls restrict physical access to computer resources and protect them from intentional or unintentional loss or impairment. Computer resources to be protected include primary computer facilities, cooling systems facilities, terminals that are used to access a computer, micro computers, computer file storage areas, and telecommunication equipment and lines, including wiring closets.

3.6.4 Visual Surveillance System: CCTV (Closed Circuit Television)

CCTV systems monitor a variety of areas. They are used as an aid to security to help control and deter acts of crime, to prevent employee pilferage, and to provide employee safety. In retail environments, they are an effective deterrent to shoplifting, vandalism, and robbery. In industrial and commercial applications, CCTV cameras are used to monitor areas that might be dangerous or inaccessible by other means.

In large complexes, CCT cameras may be located in multiple buildings and communicate to a central control point over coaxial cable, fiber, telephone wire, RF, the intranet, as well as the Internet.

3.7 BACK-UP SECURITY MEASURES

Backing up your data is an essential security measure in today's computing environment. Data has gained intrinsic value, either in the manpower needed to generate that data or in the significance of that data to your customers. While data has become more significant, it also grows, by some estimates as much as 100 percent per year. Data loss, both accidental and due to theft, costs hundreds of millions of dollars every year. When taken as a whole, one thing becomes clear: Your data must be protected.

For risks to your enterprise level, back-up security can be grouped into five areas:

3.7.1 Physical Security

- Access to servers
- Access to media libraries

- Access to media vault
- Media labels and audits

3.7.2 Client Security

Client/server communications generally follow one of the two models. In the server-based model, the server accesses the client in order to initiate the backup session. In the client-based mode, the client determines when the backups are supposed to happen and contacts the server when it is about to proceed.

- **Access to back-up client:** Anyone with access to client software has access to the data backed up by that client. Strongly consider protecting access to the client software.

3.7.3 Server Security

- Defined server access
- Encrypted data

3.7.4 Network Security

An often overlooked aspect of enterprise-level back-up security is the network. The network will carry data between client, server, and library and is potentially a huge security hole.

- **SAN security:** Storage area networking is an increasingly popular choice for providing both storage and back-up services.
- **NDMP limitations and potential:** An emerging protocol that directly affects network backups is network data management protocol or NDMP. NDMP is used primarily to provide a back-up solution for network attached storage (NAS).

3.7.5 Employee Security

- **Access to restores:** Create a file in the top directory of the backup and make the file readable only by those people allowed to request restores.
- **Back-up administrator:** Most security comes down to having responsible people. A back-up administrator is going to have access to a large amount of your data; be sure you can trust him.

EXERCISES

1. Explain briefly the process of developing secure information systems.

2. Describe in brief *application development security*.

3. What do you understand by *security architecture and design*?

4. What are security threat issues related to hardware, data storage, and downloadable devices?

5. What are the primary measures applied for the security of backup?

6. What is the process of developing a secure information system? Discuss main security considerations in developing secure information systems.

7. What is the security employed in developing applications? What are the primary issues related to the secure development of applications?

8. What are the activities involved in the process of management of risks related to information assets of an organization?

9. What are the benefits of the common framework used for developing secure applications and what are the factors of this framework? Explain design guidelines in detail.

10. What is security architecture and design? How can we design a secure system?

11. What are the security issues with storage and peripheral devices?

12. State the benefits of ISMS.

13. Define the term "SDLC." What are its phases?

14. Mention the important stages of integrating security.

15. Comment on the importance of integrating security in the implementation phase.

16. State the benefits of information security governance.

17. What are the security issues associated with storage devices?

18. Write short notes on the following:
   a. Peripheral device security
   b. CCTV
   c. Back-up security measures

19. Explain by means of a diagram the working of host IDS.

20. Define the term "disposal phase" and add a note on its security aspects.

21. Prove the hypothesis, "Systems buses are an important components in hardware security system architecture."

22. What is meant by "abstraction and authentification"?

23. How are random numbers used to develop system security?

24. What is security policy? Why and how are the policies to be framed for any organization? Is it required to review security policies once these have been framed?

25. How is the physical security ensured for IT assets? How is the access into an IT area given? Is that area vigilantly survey protected?

26. What are various hardware issues while considering security? Besides the data security, is it also necessary to have software, applications, and backup security in place?

27. Define biometric systems. Discuss the design issues in biometric systems.

CHAPTER 4

INFORMATION SECURITY POLICIES, STANDARDS, AND CYBER LAW

4.1 SECURITY POLICIES

Information security is a set of policies issued by an organization to ensure that all information technology users within the domain of the organization or its networks comply with rules and guidelines related to the security of the information stored digitally at any point in the network or within the organization's boundaries of authority.

4.1.1 Policies

An information security policy consists of *high-level statements* relating to the protection of information across the business and should be produced by senior management. Policies are long-term, high-level management instructions on how the organization is to be run and generally are driven by legal concerns (due diligence). Policies reflect an organization's goals, objectives, and culture and are intended for broad audiences. They also are mandatory and are applicable to anyone—employee, contractor, temporary, and so on. Special approval if the policy is not to be followed (an exception) should be documented. (Yes, a policy for exceptions is necessary!) Policies drive standards, procedures, and technical controls, for example, "Passwords will be used."

4.1.2 Why Policies Should be Developed

Every organization needs to protect its data and also control how it should be distributed both within and without the organization boundaries. This may mean that information may have to be encrypted, authorized through a third party or institution, and may have restrictions placed on its distribution with reference to a classification system laid out in the information security policy.

A business might employ an information security policy to protect its digital assets and intellectual rights in efforts to prevent theft of industrial secrets and information that could benefit competitors.

4.1.3 WWW Policies

The *World Wide Web* (known as *WWW, Web, W3*) is the universe of Internet-accessible information. The World Wide Web began as a networked information project at *CERN*, the European laboratory for particle physics. The Web has a body of software and a set of protocols and conventions used to traverse and find information over the Internet. Through the use of hypertext and multimedia techniques, the Web is easy for anyone to roam, browse, and contribute to.

There are a number of risks related to the use of a WWW browser to search for and retrieve information over the Internet. Web-browsing programs are very complicated and getting more complicated all the time. The more complicated a program is, the less secure it generally is. Flaws may then be network-based attacks. The general WWW policy includes the following:

a. Web browsers leave "footprints," providing a tail of all site visits.

b. Company Internet users are prohibited from transmitting or downloading material that is offensive, pornographic, threatening, or racially or sexually harassing.

c. Any files downloaded over the WWW shall be scanned for viruses, using approved virus detection software.

d. Only company-approved versions of browser software may be used or downloaded. Non-approved versions may contain viruses or other bugs.

e. URLs of offensive sites must be forwarded to the company Web security contact.

f. No offensive or harassing material may be made available via company websites.

 g. No personal commercial advertising may be made available via company websites.

 h. No company confidential material will be made available on websites.

4.1.4 Email Security Policies

 a. Do not forward any chain letter.

 b. Do not transmit unsolicited mass email (spam) to anyone.

 c. Do not send any message that supports illegal or unethical activities.

 d. Carefully handle emails.

 e. Archive emails.

 f. Scan email attachments.

 g. Limit size of an email.

 h. Encrypt emails for confidentiality.

 i. Digitally sign emails for authenticity.

4.2 POLICY REVIEW PROCESS

Information security reviews check the compliance and monitoring of security controls to detect any weaknesses that may affect the information assets and core services. There are two different forms of security reviews.

4.2.1 Technical Review

Technical reviews are carried out by computing services (CSD) technical teams on a regular basis. Regular reviews include but are not limited to the following:

 a. Networks and communication infrastructure

 b. Operating systems

 c. Middleware

 d. Databases

 e. Applications

4.2.2 Non-Technical Review

These are known as "entity-level controls" and are carried out by the information security officer to ensure that security policies, standards, processes, legislation, regulation, and best practice are adhered to. The policy review steps are the following:

a. Have someone, other than person who wrote it, review the policy.

b. Assess policy for completeness.

c. Ensure policy statements are clear, concise, and SMART (specific, measurable, achievable, realistic, and time-bound).

d. Ensure policy answers the five W's.

e. Ensure compliance to laws, regulations, and other policies.

f. Check freshness and ease of availability to organizational members.

4.2.3 Corporate Policies

A sample corporate information security policy contains the following information.

Objectives, Aim, and Scope

The objectives of an organizational information security policy are to preserve the following:

(i.) Confidentiality: Access to data should be confined to those with appropriate authority

(ii.) Integrity: Information should be complete and accurate. All systems, assets, and networks should operate correctly according to specifications.

(iii.) Availability: Information should be available and delivered to the right person, at the time when it is needed.

4.2.4 Policy Framework, Legislation, and Responsibilities of Information Security

a. Management of security

b. Security controls of assets

c. Access controls: User, computer, and application

d. Equipment security

e. Computer and network procedures

f. Information risk assessment

g. Classification of sensitive information

h. Protection of malicious software

i. Monitoring system access and use

j. System change control

k. Business continuity and disaster recovery plans

l. Intellectual property rights

4.3   INFORMATION SECURITY STANDARDS

The term "standard" is sometimes used within the context of information security policies to distinguish between written policies, standards, and procedures. Organizations should maintain all three levels of documentation to help secure their environment. A "standard" is a low-level prescription for the various ways the company will enforce a given policy (e.g., Passwords will be at least eight characters and require at least one number). While information security plays an important role in protecting the data and assets of an organization, we often hear news about security incidents, such as defacement of websites, server hacking, and data leakage. Organizations need to be fully aware of the need to devote more resources to the protection of information assets and information security must become a top concern in both government and business.

To address the situation, a number of governments and organizations have set up benchmarks, standards, and, in some cases, legal regulations on information security to help ensure an adequate level of security is maintained, resources are used in the right way, and the best security practices are adopted.

4.3.1 International Standards Organization (ISO)

The International Standards Organization (ISO), established in 1947, is a non-governmental international body that collaborates with the International Electro-technical Commission (IEC) and the International

Telecommunication Union (ITU) on information and communication technology (ICT) standards. The following are commonly referenced as ISO security standards.

4.3.1.1 ISO/IEC 27001:2005

(Information Security Management System: Requirements)

The international standard *ISO/IEC 27001:2005* has its roots in the technical content derived from *BSI standard BS7799 Part 2:2002*. It specifies the requirements for establishing, implementing, operating, monitoring, reviewing, maintaining, and improving a documented *information security management system (ISMS)* within an organization.

The standard introduces a cyclic model known as the *Plan-Do-Check Act (PDCA) model* that aims to establish, implement, monitor, and improve the effectiveness of an organization's ISMS. The PDCA cycle has following four phases:

(i.) "Plan" phase: Establish the ISMS.

(ii.) "Do" phase: Implement and operate the ISMS.

(iii.) "Check" phase: Monitor and review the ISMS.

(iv.) "Act" phase: Maintain and Improve the ISMS.

4.3.1.2 ISO/IEC 27002:2005 (Code of Practice for Information Security Management)

ISO/IEC 27002:2005 (replaced ISO/IEC 17799:2005 in April 2006) is an international standard that originated from *BS7799 – 1*, one that was originally laid down by the *British Standards Institute (BSI)*. ISO/IEC 27002: 2005 refers to a code of practice for information security management and is intended as a common basis and practical guideline for developing organizational security standards and effective management practices.

This standard contains guidelines and best practices recommendations for these ten security domains:

a. Security policy

b. Organization of information security

c. Asset management

d. Human resources security

e. Physical and environmental security

f. Communications and operations management

g. Access control

h. Information system acquisition, development, and maintenance

i. Information security incident management

j. Business continuity management

k. Compliance

Among these ten security domains, a total of 39 control objectives and hundreds of best-practice information security control measures are recommended for organizations to satisfy the control objectives and protect information assets against threats to confidentiality, integrity, and availability.

4.3.1.3 ISO/IEC 15408 (Evaluation Criteria for IT Security)

The international standard ISO/IEC 15408 is commonly known as the "Common Crieteria" (CC) 12. It consists of three parts:

a. *ISO/IEC 15408 – 1:2005* (introduction and general model)

b. *ISO/IEC 15408 – 2:2005* (security functional requirements)

c. *ISO/IEC 15408 – 3:2005* (security assurance requirements)

4.3.1.4 ISO/IEC 13335 (IT Security Management)

ISO/IEC13335, initially a technical report (TR) before becoming a full ISO/IEC standard, consists of a series of guidelines for technical security control measures:

a. *ISO/IEC 13335 – 1:2004* documents the concepts and models for information and communication technology security management.

b. *ISO/IEC TR 13335 – 3:1998* documents the techniques for management of IT security. This is under review and may be superseded by *ISO/IEC 27005*.

c. *ISO/IEC TR 13335 – 4:2000* covers the selection of safeguards (i.e., technical security controls). This is under review and may be superseded by *ISO/IEC 27005*.

d. *ISO/IEC TR 13335 – 5:2001* covers management guidance on network security. This is also under review and may be merged into *ISO/IEC 18028 – 1 and ISO/IEC 27033*.

4.4 CYBER LAWS IN INDIA

In a cybercrime, a computer or data itself is the target or the objective of offense, or a tool in committing some other offense, providing the necessary inputs for that offense. All such acts of crime will come under the broader definition of cybercrime.

The mid-90s saw an impetus in globalization and computerization, with more and more nations computerizing their governance and e-commerce seeing enormous growth. Until then, most of international trade and transactions were done through documents being transmitted through post and telex only. Evidence and records, until then, were predominantly paper or an other forms of hard copy only. With much of international trade being done through electronic communication and with email gaining momentum, an urgent and imminent need was felt for recognizing electronic records (i.e., the data stored in a computer or an external storage attached thereto).

The *United Nations Commission on International Trade Law (UNICTRAL)* adopted the *Model Law* on e-commerce in 1996. The *General Assembly of United Nations* passed a resolution in January 1997 inter alia, recommending all states in the UN give favorable considerations to said Model Law, which provides for recognition to electronic records, and give it the same treatment as paper communication and records. It is against this background that the government of India enacted its *Information Technology Act of 2000* with the objectives as follows, stated in the preface to the act itself: "to provide legal recognition for transactions carried out by means of electronic data interchange and other means of electronic communication, commonly referred to as "electronic commerce," which involves the use of alternatives to paper-based methods of communication and stage of information, to facilitate electronic filling of documents with the government agencies and further to amend the Indian Penal Code, the Indian Evidence Act 1872, the Banker's Books Evidence Act 1891 and the Reserve Bank of India Act, 1934 and for matters connected therewith or incidental thereto."

The Information Technology Act of 2000, was thus passed as the Act No. 21 of 2000.

The president's ascent on June 9, 2000 and was made effective from October 17, 2000.

The act essentially deals with following issues:

a. Legal recognition of electronic documents

b. Legal recognition of digital signature

    c. Offenses and contraventions

    d. Justice dispensation systems for cybercrimes

4.4.1 Amendment of Act 2008

Being the first legislation in the nation on technology, computers, and e-commerce and e-communication, the act was the subject of extensive debates, elaborate reviews, and detailed criticisms, with one arm of the industry criticizing some sections of the act to be draconian and others stating it is too diluted and lenient. The Indian Penal Code, even in technology-based cases with the IT. Act relies more on IPC rather on ITA.

    Some of the notable features of ITAA are as follows:

    a. Focusing on data privacy

    b. Focusing on information security

    c. Defining cyber café

    d. Making digital signature technology neutral

    e. Defining reasonable security practices to be followed by corporate

    f. Redefining the role of intermediaries

    g. Recognizing the role of the Indian Compute Emergency Response Team

    h. Inclusion of some additional cybercrimes like child pornography and cyberterrorism

    i. Authorizing an inspector to investigate cyber offenses (as against the DSP earlier)

4.4.2 IT Act of 2000 Provisions

The Information Technology Act of 2000 consisted of ninety-four sections segregated into thirteen chapters. Four schedules form part of the act. In the 2008 versions of the act, there are 124 sections (excluding five sections that have been omitted from the earlier version) and fourteen chapters. Schedule I and II have been replaced. Schedules III and IV are deleted.

    The Information Technology Act of 2000 addressed the following issues:

    a. Legal recognition of electronic documents

    b. Legal recognition of digital signatures

c. Offenses and contraventions

d. Justice dispensation systems for cybercrimes

Chapter number	Provisions under the chapter
Chapter 1	Preliminary definitions
Chapter 2	Digital signature
Chapter 3	Electronic governance
Chapter 4	Attribution, acknowledgement, and dispatch of electronic records
Chapter 5	Secure electronic records and secure digital signatures
Chapter 6	Regulation of certifying authorities
Chapter 7	Digital signature certificates
Chapter 8	Duties of subscribers
Chapter 9	Penalties and adjudication
Chapter 10	The Cyber Regulations Appellate Tribunal
Others	• Offenses • Network service providers not liable in certain cases

This act was amended by the *Information Technology Amendment Bill of 2006*, passed in Lok Sabha on December 22, 2000 and in Rajya Sabha on December 23, 2000.

The amended act was to provide legal recognition for the transactions carried out by means of electronic data interchange and other means of electronic communication, commonly referred to as "electronic commerce," which involves the use of alternatives to paper-based methods of communication and storage of information, to facilitate electronic filings of documents with the government agencies, and further to amend the *Indian Penal Code, Indian Evidence Act of 1872*, the *Banker's Books Evidence Act of 1891 and the Reserve Bank of India Act of 1983* and for matters connected therewith or incidental thereto.

4.5 INTELLECTUAL PROPERTY LAW

Intellectual property (IP) is a term referring to creations of the intellect for which a monopoly is assigned to designated owners by law. Some common types of intellectual property rights (IPR) are *trademarks, copyright, patents, industrial design rights*, and in some jurisdictions trade secrets; all these cover music, literature, and other artistic works; discoveries and inventions; and words, phrases, symbols, and designs.

While intellectual property law has evolved over centuries, it was not until the nineteenth century that the term "intellectual property" began to be used, and not until the late twentieth century that it became commonplace in the majority of the world.

4.5.1 History

The Statute of Monopolies (1624) and the British Statute of Anne (1710) are seen as the origins of patent law and copyright respectively, firmly establishing the concept of intellectual property.

The first known use of the term "intellectual property" dates to 1769, when a piece published in the *Monthly Review* used the phrase. The first clear example of modern usage goes back as early as 1808, when it was used as a heading title in a collection of essays.

The German equivalent was used with the founding of the North German Confederation, whose constitution granted legislative power over the protection of intellectual property (Schutz des geistigen Eigentums) to the confederation. When the administrative secretariats established by the Paris Convention (1883) and the Berne Convention (1886) merged in 1893, they located in Berne, and also adopted the term "intellectual property" in their new combined title, the United International Bureaux for the Protection of Intellectual Property.

The organization subsequently relocated to Geneva in 1960 and was succeeded in 1967 with the establishment of the World Intellectual Property Organization (WIPO) by treaty as an agency of the United Nations. According to Lemley, it was only at this point that the term really began to be used in the United States (which had not been a party to the Berne Convention), and it did not enter popular usage until passage of the Bayh-Dole Act in 1980.

"The history of patents does not begin with inventions, but rather with royal grants by Queen Elizabeth I (1558–1603) for monopoly privileges. . . . Approximately 200 years after the end of Elizabeth's reign, however, a patent represents a legal right obtained by an inventor providing for exclusive control over the production and sale of his mechanical or scientific invention . . . [demonstrating] the evolution of patents from royal prerogative to common-law doctrine."

The term can be found in an October 1845 Massachusetts Circuit Court ruling in the patent case *Davoll et al. v. Brown*, in which Justice Charles L. Woodbury wrote that "only in this way can we protect intellectual property, the labors of the mind, productions and interests that are as much a man's own . . . as the wheat he cultivates, or the flocks he rears." The statement that "discoveries are . . . property" goes back earlier. Section 1 of the French law of 1791 stated, "All new discoveries are the property of the author; to assure the inventor the property and temporary enjoyment of his discovery, there shall be delivered to him a patent for five, ten, or fifteen years." In Europe, French author A. Nion mentioned *propriété intellectuelle* in his *"Droits Civils des Auteurs, Artistes et Inventeurs,"* published in 1846.

Until recently, the purpose of intellectual property law was to give as little protection possible in order to encourage innovation. Historically, therefore, they were granted only when they were necessary to encourage invention, limited in time and scope.

The concept's origins can potentially be traced back further. Jewish law includes several considerations whose effects are similar to those of modern intellectual property laws, though the notion of intellectual creations as property does not seem to exist; notably, the principle of Hasagat Ge'vul (unfair encroachment) was used to justify limited-term publisher (but not author) copyright in the sixteenth century. In 500 BCE, the government of the Greek state of Sybaris offered one year's patent "to all who should discover any new refinement in luxury."

Intellectual Property Rights

Intellectual property rights include patents, copyright, industrial design rights, trademarks, plant variety rights, trade dress, and, in some jurisdictions, trade secrets. There are also more specialized or derived varieties of *sui generis* exclusive rights, such as circuit design rights (called "mask work rights" in the US) and supplementary protection certificates for pharmaceutical products (after expiration of a patent protecting them) and database rights (in European law).

Patents

A patent is a form of right granted by the government to an inventor, giving the owner the right to exclude others from making, using, selling, offering to sell, and importing an invention for a limited period of time, in exchange for the public disclosure of the invention. An invention is a solution to a specific technological problem, which may be a product or a process and generally has to fulfill three main requirements: It has to be new, not obvious, and there needs to be an industrial applicability.

Copyright

A copyright gives the creator of an original work exclusive rights to it, usually for a limited time. Copyright may apply to a wide range of creative, intellectual, or artistic forms, or "works." Copyright does not cover ideas and information themselves, only the form or manner in which they are expressed.

4.5.2 Industrial Design Rights

An industrial design right (sometimes called "design right") protects the visual design of objects that are not purely utilitarian. An industrial design consists of the creation of a shape, configuration or composition of pattern or color, or combination of pattern and color in three-dimensional form containing aesthetic value. An industrial design can be a two- or three-dimensional pattern used to produce a product, industrial commodity, or handicraft.

Plant Varieties

Plant breeders' rights or plant variety rights are the rights to commercially use a new variety of a plant. The variety must, among others, be novel and distinct, and for registration the evaluation of propagating material of the variety is examined.

Trademarks

A trademark is a recognizable sign, design, or expression that distinguishes products or services of a particular trader from the similar products or services of other traders.

Trade Dress

Trade dress is a legal term of art that generally refers to characteristics of the visual appearance of a product or its packaging (or even the design of a building) that signifies the source of the product to consumers.

Trade Secrets

A trade secret is a formula, practice, process, design, instrument, pattern, or compilation of information that is not generally known or reasonably ascertainable, by which a business can obtain an economic advantage over competitors or customers.

4.5.3 Objectives of Intellectual Property Law

The stated objective of most intellectual property law (with the exception of trademarks) is to "promote progress." By exchanging limited exclusive rights for disclosure of inventions and creative works, society and the patentee/copyright owner mutually benefit, and an incentive is created for inventors and authors to create and disclose their work. Some commentators have noted that the objective of intellectual property legislators and those who support its implementation appears to be "absolute protection." "If some intellectual property is desirable because it encourages innovation, they reason, more is better. The thinking is that creators will not have sufficient incentive to invent unless they are legally entitled to capture the full social value of their inventions." This absolute protection or full value view treats intellectual property as another type of "real" property, typically adopting its law and rhetoric. Other recent developments in intellectual property law, such as the America Invents Act, stress international harmonization. Recently there has also been much debate over the desirability of using intellectual property rights to protect cultural heritage, including intangible ones, as well as over risks of commodification derived from this possibility. The issue still remains open in legal scholarship.

1. Financial Incentive

These exclusive rights allow owners of intellectual property to benefit from the property they have created, providing a financial incentive for the creation of an investment in intellectual property, and, in case of patents, pay associated research and development costs. Some commentators, such as David Levine and Michele Boldrin, dispute this justification.

In 2013, the United States Patent and Trademark Office approximated that the worth of intellectual property to the US economy is more than $5 trillion and creates employment for an estimated 18 million American people. The value of intellectual property is considered similarly high in other developed nations, such as those in the European Union. In the UK, IP has become a recognized asset class for use in pension-led funding and other types of business finance. However, in 2013, the UK Intellectual Property Office stated, "There are millions of intangible business assets whose value is either not being leveraged at all, or only being leveraged inadvertently."

2. Economic Growth

The WIPO treaty and several related international agreements underline that the protection of intellectual property rights is essential to maintaining economic growth. The *WIPO Intellectual Property Handbook* gives two reasons for intellectual property laws.

One is to give statutory expression to the moral and economic rights of creators in their creations and the rights of the public in access to those creations. The second is to promote, as a deliberate act of government policy, creativity and the dissemination and application of its results and to encourage fair trading, which would contribute to economic and social development.

The Anti-Counterfeiting Trade Agreement (ACTA) states that "effective enforcement of intellectual property rights is critical to sustaining economic growth across all industries and globally."

Economists estimate that two-thirds of the value of large businesses in the United States can be traced to intangible assets. "IP-intensive industries" are estimated to generate 72 percent more value added (price minus material cost) per employee than "non-IP-intensive industries."

A joint research project of the WIPO and the United Nations University measuring the impact of IP systems on six Asian countries found "a positive correlation between the strengthening of the IP system and subsequent economic growth."

Economists have also shown that IP can be a disincentive to innovation when that innovation is drastic. IP makes excludable non-rival intellectual products that were previously non-excludable. This creates economic inefficiency as long as the monopoly is held. A disincentive to direct resources toward innovation can occur when monopoly profits are less than the overall welfare improvement to society. This situation can be seen as a market failure and an issue of appropriability.

3. Morality

According to Article 27 of the Universal Declaration of Human Rights, "Everyone has the right to the protection of the moral and material interests resulting from any scientific, literary or artistic production of which he is the author." Although the relationship between intellectual property and human rights is a complex one, there are moral arguments for intellectual property.

The arguments that justify intellectual property fall into three major categories. Personality theorists believe intellectual property is an extension of an individual. Utilitarians believe that intellectual property stimulates social progress and pushes people to further innovation. Lockeans argue that intellectual property is justified based on deservedness and hard work.

Various moral justifications for private property can be used to argue in favor of the morality of intellectual property, such as the following:

a. **Natural rights/justice argument:** This argument is based on Locke's idea that a person has a natural right over the labor and/or products produced by his or her body. Appropriating these products is viewed as unjust. Although Locke had never explicitly stated that natural right applied to products of the mind, it is possible to apply his argument to intellectual property rights, in which it would be unjust for people to misuse another's ideas. Locke's argument for intellectual property is based on the idea that laborers have the right to control that which they create. They argue that we own our bodies, which are the laborers, and this right of ownership extends to what we create. Thus, intellectual property ensures this right when it comes to production.

b. **Utilitarian/pragmatic argument:** According to this rationale, a society that protects private property is more effective and prosperous than societies that do not. Innovation and invention in nineteenth-century America has been attributed to the development of the patent system. By providing innovators with "durable and tangible return on their investment of time, labor, and other resources," intellectual property rights seek to maximize social utility. The presumption is that they promote public welfare by encouraging the "creation, production, and distribution of intellectual works." Utilitarians argue that without intellectual property there would be a lack of incentive to produce new ideas. Systems of protection such as intellectual property optimize social utility.

c. **"Personality" argument:** This argument is based on a quote from Hegel: "Every man has the right to turn his will upon a thing or make the thing an object of his will, that is to say, to set aside the mere thing and recreate it as his own." European intellectual property law is shaped by this notion that ideas are an "extension of oneself and of one's personality." Personality theorists argue that by being a creator of something one is inherently at risk and vulnerable for having their ideas and designs stolen and/or altered. Intellectual property protects these moral claims that have to do with personality.

Lysander Spooner argues "that a man has a natural and absolute right—and if a natural and absolute, then necessarily a perpetual, right—of property, in the ideas, of which he is the discoverer or creator; that his right of property, in ideas, is intrinsically the same as, and stands on identically the same grounds with, his right of property in material things; that no distinction, of principle, exists between the two cases."

Writer Ayn Rand argued in her book *Capitalism: The Unknown Ideal* that the protection of intellectual property is essentially a moral issue. The belief is that the human mind itself is the source of wealth and survival and that all property at its base is intellectual property. To violate intellectual property is therefore no different morally than violating other property rights, which compromises the very processes of survival and therefore constitutes an immoral act.

4.5.4 Infringement, Misappropriation, and Enforcement

Violation of intellectual property rights, called "infringement," with respect to patents, copyright, and trademarks, and "misappropriation" with respect to trade secrets, may be a breach of civil law or criminal law, depending on the type of intellectual property involved, jurisdiction, and the nature of the action.

As of 2011, trade in counterfeit copyrighted and trademarked works was a $600 billion industry worldwide and accounted for 5–7 percent of global trade.

4.5.4.1 Patent Infringement

Patent infringement typically is caused by using or selling a patented invention without permission from the patent holder. The scope of the patented invention or the extent of protection is defined in the claims of the granted patent. There is safe harbor in many jurisdictions to use a patented invention for research. This safe harbor does not exist in the US unless the research is done for purely philosophical purposes, or in order to gather data in order to prepare an application for regulatory approval of a drug. In general, patent infringement cases are handled under civil law (e.g., in the United States) but several jurisdictions incorporate infringement in criminal law also (for example, Argentina, China, France, Japan, Russia, and South Korea).

4.5.4.2 Copyright Infringement

Copyright infringement is reproducing, distributing, displaying, or performing a work, or making derivative works, without permission from the copyright holder, which is typically a publisher or other business represented or assigned by the work's creator. It is often called "piracy." While copyright is created the instance a work is fixed, generally the copyright holder can only get money damages if the owner registers the copyright. Enforcement of copyright is generally the responsibility of the copyright holder. The ACTA trade agreement, signed in May 2011 by the United States, Japan, Switzerland, and the EU, and which has not entered into force, requires that its parties add criminal penalties, including incarceration and fines, for copyright and trademark infringement, and obligated the parties to active police for infringement. There are limitations and exceptions to copyright, allowing limited use of copyrighted works, which does not constitute infringement. Examples of such doctrines are the fair use and fair dealing doctrine.

4.5.4.3 Trademark Infringement

Trademark infringement occurs when one party uses a trademark that is identical or confusingly similar to a trademark owned by another party, in relation to products or services that are identical or similar to the products or services of the other party. In many countries, a trademark receives protection without registration, but registering a trademark provides legal advantages for enforcement. Infringement can be addressed by civil litigation and, in several jurisdictions, under criminal law.

4.5.4.4 Trade Secret Misappropriation

Trade secret misappropriation is different from violations of other intellectual property laws, since by definition trade secrets are secret, while patents and registered copyrights and trademarks are publicly available. In the United States, trade secrets are protected under state law, and states have nearly universally adopted the Uniform Trade Secrets Act. The United States also has federal law in the form of the Economic Espionage Act of 1996 (18 U.S.C. §§ 1831–1839), which makes the theft or misappropriation of a trade secret a federal crime. This law contains two provisions criminalizing two sorts of activity. The first, 18 U.S.C. § 1831(a), criminalizes the theft of trade secrets to benefit foreign powers. The second, 18 U.S.C. § 1832, criminalizes theft for commercial or economic purposes. (The statutory penalties are different for the two offenses.) In commonwealth common law jurisdictions, confidentiality and trade secrets are regarded as an equitable right rather than a property right, but penalties for theft are roughly the same as the United States.

4.5.4.5 Criticisms

Criticism of the term "intellectual property" ranges from discussing its vagueness and abstract overreach to direct contention to the semantic validity of using words like "property" and "rights" in fashions that contradict practice and law. Many detractors think this term specially serves the doctrinal agenda of parties opposing reform in the public interest or otherwise abusing related legislations and that it disallows intelligent discussion about specific and often unrelated aspects of copyright, patents, trademarks, and so on.

Free Software Foundation founder Richard Stallman argues that, although the term intellectual property is in wide use, it should be rejected altogether, because it "systematically distorts and confuses these issues, and its use was and is promoted by those who gain from this confusion." He claims that the term "operates as a catch-all to lump together disparate laws [which] originated separately, evolved differently, cover different activities, have different rules, and raise different public policy issues" and that it creates a "bias" by confusing these monopolies with ownership of limited physical things, likening them to "property rights." Stallman advocates referring to copyrights, patents, and trademarks in the singular and warns against abstracting disparate laws into a collective term.

Similarly, economists Boldrin and Levine prefer to use the term "intellectual monopoly" as a more appropriate and clear definition of the concept, which, they argue, is very dissimilar from property rights.

On the assumption that intellectual property rights are actual rights, Stallman argues that this claim does not live to the historical intentions behind these laws, which in the case of copyright served as a censorship system, and later on, a regulatory model for the printing press that may have benefited authors incidentally, but never interfered with the freedom of average readers. Still referring to copyright, he cites legal literature such as the United States Constitution and case law to demonstrate that it is meant to be an optional and experimental bargain that temporarily trades property rights and free speech for public, not private, benefit in the form of increased artistic production and knowledge. He mentions that "if copyright were a natural right nothing could justify terminating this right after a certain period of time."

Law professor, writer, and political activist Lawrence Lessig, along with many other copy and free software activists, has criticized the implied analogy with physical property (like land or an automobile). He argues such an analogy fails because physical property is generally rivalrous while intellectual works are non-rivalrous (that is, if one makes a copy of a work, the enjoyment of the copy does not prevent enjoyment of the original). Other arguments along these lines claim that unlike the situation with tangible property, there is no natural scarcity of a particular idea or information: Once it exists at all, it can be reused and duplicated indefinitely without such reuse diminishing the original. Stephan Kinsella has objected to intellectual property on the grounds that the word "property" implies scarcity, which may not be applicable to ideas.

Entrepreneur and politician Rickard Falkvinge and hacker Alexandre Oliva have independently compared George Orwell's fictional dialect Newspeak to the terminology used by intellectual property supporters as a linguistic weapon to shape public opinion regarding copyright debate and DRM.

4.5.4.6 Alternative Terms

In civil law jurisdictions, intellectual property has often been referred to as intellectual rights, traditionally a somewhat broader concept that has included moral rights and other personal protections that cannot be bought or sold. Use of the term "intellectual rights" has declined since the early 1980s, as use of the term "intellectual property" has increased.

The alternative terms "monopolies on information" and "intellectual monopoly" have emerged among those who argue against the "property or "intellect" or "rights" assumptions, notably Richard Stallman. The backronyms intellectual protectionism and intellectual poverty, whose initials are also IP, have found supporters as well, especially among those who have used the backronym digital restrictions management.

The argument that an intellectual property right should (in the interests of better balancing of relevant private and public interests) be termed an intellectual monopoly privilege (IMP) has been advanced by several academics, including Birgitte Andersen and Thomas Alured Faunce.

4.5.4.7 Objections to Overbroad Intellectual Property Laws

Some critics of intellectual property, such as those in the free culture movement, point at intellectual monopolies as harming health (in the case of pharmaceutical patents), preventing progress, and benefiting concentrated interests to the detriment of the masses and argue that the public interest is harmed by ever-expansive monopolies in the form of copyright extensions, software patents, and business method patents. More recently scientists and engineers are expressing concern that patent thickets are undermining technological development even in high-tech fields like nanotechnology.

Petra Moser has asserted that historical analysis suggests that intellectual property laws may harm innovation.

Overall, the weight of the existing historical evidence suggests that patent policies, which grant strong intellectual property rights to early generations of inventors, may discourage innovation. On the contrary, policies that encourage the diffusion of ideas and modify patent laws to facilitate entry and encourage competition may be an effective mechanism to encourage innovation.

Peter Drahos notes, "Property rights confer authority over resources.

When authority is granted to the few over resources on which many depend, the few gain power over the goals of the many. This has consequences for both political and economic freedoms with in a society."

The World Intellectual Property Organization (WIPO) recognizes that conflicts may exist between the respect for and implementation of current intellectual property systems and other human rights. In 2001 the UN Committee on Economic, Social, and Cultural Rights issued a document called "Human Rights and Intellectual Property" that argued that intellectual property tends to be governed by economic goals when it should be viewed primarily as a social product; in order to serve human well-being, intellectual property systems must respect and conform to human rights laws. According to the committee, when systems fail to do so they risk infringing upon the human right to food and health, and to cultural participation and scientific benefits. In 2004, the General Assembly of WIPO adopted the Geneva Declaration on the Future of the World Intellectual Property Organization, which argues that WIPO should "focus more on the needs of developing countries, and to view IP as one of many tools for development—not as an end in itself."

Further along these lines, the ethical problems brought up by IP rights are most pertinent when socially valuable goods like life-saving medicines are given IP protection. While the application of IP rights can allow companies to charge more than the marginal cost of production in order to recoup the costs of research and development, the price may exclude from the market anyone who cannot afford the cost of the product, in this case a life-saving drug. "An IPR driven regime is therefore not a regime that is conductive to the investment of R&D of products that are socially valuable to predominately poor populations."

Some libertarian critics of intellectual property have argued that allowing property rights in ideas and information creates artificial scarcity and infringes on the right to own tangible property. Stephan Kinsella uses the following scenario to argue this point:

> Imagine the time when men lived in caves. One bright guy—let's call him Galt-Magnon—decides to build a log cabin on an open field, near his crops. To be sure, this is a good idea, and others notice it. They naturally imitate Galt-Magnon, and they start building their own cabins. But the first man to invent a house, according to IP advocates, would have a right to prevent others from building houses on their own land, with their own logs, or to charge them a fee if they do build houses. It is plain that the innovator in these examples becomes a partial owner of the tangible property (e.g., land

and logs) of others, due not to first occupation and use of that property (for it is already owned), but due to his coming up with an idea. Clearly, this rule flies in the face of the first-user homesteading rule, arbitrarily and groundlessly overriding the very homesteading rule that is at the foundation of all property rights.

Thomas Jefferson once said in a letter to Isaac McPherson on August 13, 1813,

If nature has made any one thing less susceptible than all others of exclusive property, it is the action of the thinking power called an idea, which an individual may exclusively possess as long as he keeps it to himself; but the moment it is divulged, it forces itself into the possession of every one, and the receiver cannot dispossess himself of it. Its peculiar character, too, is that no one possesses the less, because every other possesses the whole of it. He who receives an idea from me, receives instruction himself without lessening mine; as he who lights his taper at mine, receives light without darkening me.

In 2005, the RSA launched the Adelphi Charter, aimed at creating an international policy statement to frame how governments should make balanced intellectual property law.

Another limitation of current US Intellectual Property legislation is its focus on individual and joint works; thus, copyright protection can only be obtained in "original" works of authorship. This definition excludes any works that are the result of community creativity, for example Native American songs and stories; current legislation does not recognize the uniqueness of indigenous cultural "property" and its ever-changing nature. Simply asking native cultures to "write down" their cultural artifacts on tangible mediums ignores their necessary orality and enforces a Western bias of the written form as more authoritative.

4.5.4.8 Expansion in Nature and Scope of Intellectual Property Laws

Other criticism of intellectual property law concerns the expansion of intellectual property, both in duration and in scope.

In addition, as scientific knowledge has expanded and allowed new industries to arise in fields such as biotechnology and nanotechnology, originators of technology have sought IP protection for the new technologies. Patents have been granted for living organisms (and in the United States, certain living organisms have been patentable for over a century).

The increase in terms of protection is particularly seen in relation to copyright, which has recently been the subject of serial extensions in the United States and in Europe. With no need for registration or copyright notices, this is thought to have led to an increase in orphan works (copyrighted works for which the copyright owner cannot be contacted), a problem that has been noticed and addressed by governmental bodies around the world.

4.6   SEMICONDUCTOR LAW

The *Semiconductor Chip Protection Act of 1984 (or SCPA)* is an act of US Congress that makes the layouts of integrated circuits legally protected upon registration; hence, it is illegal to copy without permission.

4.6.1 Background

Prior to 1984, it was not necessarily illegal to produce a competing chip with an identical layout. As the legislative history for the SCPA explained, patent and copyright protection for chip layouts, chiptopographies, was largely unavailable. This led to considerable complaint by US chip manufacturers—notably, Intel, which, along with the Semiconductor Industry Association (SIA), took the lead in seeking remedial legislation—against what they termed "chip piracy." During the hearings that led to enactment of the SCPA, chip industry representatives asserted that a pirate could, for $10,000, copy a chip design that had cost its original manufacturer upwards from $100,000 to design.

4.6.2 Enactment of US and Other National Legislation

In 1984, the United States enacted the Semiconductor Chip Protection Act of 1984 (the SCPA) to protect the topography of semiconductor chips. The SCPA is found in Title 17, U.S. Code, sections 901–914.

Japan and European Community (EC) countries soon followed suit and enacted their own, similar laws protecting the topography of semiconductor chips.

Chip topographies are also protected by TRIPS, an international treaty.

4.6.3 How the SCPA Operates

Sui Generis Law

Although the US SCPA is codified in Title 17 (copyrights), the SCPA is not a copyright or patent law. Rather, it is a *sui generis* law resembling a utility model law or Gebrauchsmuster. It has some aspects of copyright law, some aspects of patent law, and in some ways, it is completely different from either.

The Semiconductor Chip Protection Act of 1984 was an innovative solution to this new problem of technology-based industry. While some copyright principles underlie the law, as do some attributes of patent law, the act was uniquely adapted to semiconductor mask works in order to achieve appropriate protection for original designs while meeting the competitive needs of the industry and serving the public interest."

In general, the chip topography laws of other nations are also *sui generis* laws. Nevertheless, copyright and patent case law illuminate many aspects of the SCPA and its interpretation.

4.6.4 Acquisition of Protection by Registration

Chip protection is acquired under the SCPA by filing with the US Copyright Office an application for "mask work" registration under the SCPA, together with a filing fee. The application must be accompanied by identifying material, such as pictorial representations of the IC layers so that in the event of infringement litigation, it can be determined what the registration covers. Protection continues for ten years from the date of registration.

4.6.5 Mask Works

The SCPA repeatedly refers to "mask works." The term is a relic of the original form of the bill that became the SCPA and was passed in the Senate as an amendment to the Copyright Act. The term "mask work" is parallel to and consistent with the terminology of the 1976 Copyright Act, which introduced the concept of "literary works," "pictorial works," "audiovisual works," and the like and protected physical embodiments of such works, such as books, paintings, video game cassettes, and the like against unauthorized copying and distribution. The terminology became unnecessary when the House of Representatives insisted on the substitution of a *sui generis* bill, but the SCPA as enacted still continued its use. The term "mask work" is not limited to actual masks used in chip manufacturing but is defined broadly in the SCPA to include the topographic creation embodied in the masks and chips. Moreover, the SCPA protects any physical embodiment of a mask work.

4.6.6 Enforcement

The owner of mask work rights may pursue an alleged infringer ("chip pirate") by bringing an action for mask work infringement in federal district court. The remedies available correspond generally to those of copyright law and patent law.

4.6.7 Functionality Unprotected

The SCPA does not protect functional aspects of chip designs, which is reserved to patent law. Although EPROM and other memory chip topographies are protectable under the SCPA, such protection does not extend to the information stored in chips, such as computer programs. Such information is protected, if at all, only by copyright law.

4.6.8 Reverse Engineering Allowed

The SCPA permits competitive emulation of a chip by means of reverse engineering. The ordinary test for illegal copying (mask work infringement) is the "substantial similarity" test of copyright law, but when the defense of reverse engineering is involved and supported by probative evidence (usually, the so-called paper trail of design and development work), the similarity must be greater. Then, the accused chip topography must be substantially identical (truly copied by rote, so-called slavish copying) rather than just substantially similar for the defendant to be liable for infringement. Most world chip topography protection laws provide for a reverse engineering privilege.

4.7   SOFTWARE LICENSES

A *software license* is a legal instrument (usually by way of contract law, with or without printed material) governing the use or redistribution of software. Under United States copyright law, all software is copyright protected, in source code and also object code form. The only exception is software in the public domain. A typical software license grants the licensee, typically an end user, permission to use one or more copies of software in ways where such a use would otherwise potentially constitute copyright infringement of the software owner's exclusive rights under copyright law.

4.7.1 Software Licenses and Copyright Law

Two common categories for software under copyright law, and therefore for licenses that grant the licensee specific rights, are proprietary software and free and open source software (FOSS). The distinct conceptual difference between the two is the granting of rights to modify and reuse a software product obtained by a customer: FOSS software licenses both rights to the customer and therefore bundles the modifiable source code with the software ("open source"), while proprietary software typically does not license these rights and therefore keeps the source code hidden ("closed source").

In addition to granting rights and imposing restrictions on the use of copyrighted software, software licenses typically contain provisions that allocate liability and responsibility between the parties entering into the license agreement. In enterprise and commercial software transactions, these terms often include limitations of liability, warranties and warranty disclaimers, and indemnity if the software infringes intellectual property rights of others.

Unlicensed software outside the copyright protection is either public domain software (PD) or software that is non-distributed, non-licensed, and handled as an internal business trade secret. Contrary to popular belief, distributed unlicensed software (not in the public domain) is fully copyright protected, and therefore legally unusable (as no usage rights at all are granted by a license) until it passes into public domain after the copyright term.[3] Examples for this are unauthorized software leaks or software projects that are placed on public software repositories like GitHub without a specified license. As voluntarily handing software into the public domain (before reaching the copyright term) is problematic in some international law domains (for instance the law of Germany), there are also licenses granting PD-like rights, for instance the CC0 or WTFPL.

Software licenses and rights are granted in context of the copyright according to Mark Webbink and are expanded by freeware and sublicensing.

Rights granted	Public domain	Permissive FOSS license (e.g., BSD license)	Copyleft FOSS license (e.g., GPL)	Freeware/ share ware/ freemium	Proprietary license	Trade secret
Copy right retained	No	Yes	Yes	Yes	Yes	Yes
Right to perform	Yes	Yes	Yes	Yes	Yes	No

(*continued*)

Rights granted	Public domain	Permissive FOSS license (e.g., BSD license)	Copyleft FOSS license (e.g., GPL)	Freeware/ share ware/ freemium	Proprietary license	Trade secret
Right to display	Yes	Yes	Yes	Yes	Yes	No
Right to copy	Yes	Yes	Yes	Often	No	No
Right to modify	Yes	Yes	Yes	No	No	No
Right to distribute	Yes	Yes, under same license	Yes, under same license	Often	No	No
Right to sublicense	Yes	Yes	No	No	No	No
Example software	SQ Lite, image	Apache Web server, Toy Box	Linux kernel, GIMP	Irfan view, Winamp	Windows, Half-Life 2	Server-side World of Warcraft

4.7.2 Ownership versus Licensing

In the United States, Section 117 of the Copyright Act gives the owner of a particular copy of software the explicit right to use the software with a computer, even if use of the software with a computer requires the making of incidental copies or adaptations (acts which could otherwise potentially constitute copyright infringement). Therefore, the owner of a copy of computer software is legally entitled to use that copy of software. Hence, if the end user of software is the owner of the respective copy, then the end user may legally use the software without a license from the software publisher.

As many proprietary "licenses" only enumerate the rights that the user already has under 17 U.S.C. § 117 and yet proclaim to take rights away from the user, these contracts may lack consideration. Proprietary software licenses often proclaim to give software publishers more control over the way their software is used by keeping ownership of each copy of software with the software publisher. By doing so, Section 117 does not apply to the end user and the software publisher may then compel the end user to accept all of the terms of the license agreement, many of which may be more restrictive than copyright law alone. The form of the relationship determines if it is a lease or a purchase (for example *UMG v. Augusto* or *Vernor v. Autodesk, Inc.*).

The ownership of digital goods, like software applications and video games, is challenged by "licensed, not sold" EULAs of digital distributors like Steam. In the European Union, the European Court of Justice held that a copyright holder cannot oppose the resale of a digitally sold software, in accordance with the rule of copyright exhaustion on first sale as ownership is transferred, and questions therefore the "licensed, not sold" EULA. The Swiss-based company UsedSoft innovated the resale of business software and fought for this right in court.

4.7.3 Proprietary Software Licenses

The hallmark of proprietary software licenses is that the software publisher grants the use of one or more copies of software under the end-user license agreement (EULA), but ownership of those copies remains with the software publisher (hence use of the term "proprietary"). This feature of proprietary software licenses means that certain rights regarding the software are reserved by the software publisher. Therefore, it is typical of EULAs to include terms that define the uses of the software, such as the number of installations allowed or the terms of distribution. The most significant effect of this form of licensing is that if ownership of the software remains with the software publisher, then the end user must accept the software license. In other words, without acceptance of the license, the end user may not use the software at all. One example of such a proprietary software license is the license for Microsoft Windows. As is usually the case with proprietary software licenses, this license contains an extensive list of activities that are restricted, such as reverse engineering, simultaneous use of the software by multiple users, and publication of benchmarks or performance tests.

The most common licensing models are per single user (named user, client, node) or per user in the appropriate volume discount level, while some manufacturers accumulate existing licenses. These open volume license programs are typically called open license programs (OLP), transactional license programs (TLP), volume license programs (VLP) and so on, and are contrary to the contractual license program (CLP), where the customer commits to purchase a certain number of licenses over a fixed period (mostly two years). Licensing per concurrent/floating user also occurs, where all users in a network have access to the program, but only a specific number at the same time. Another license model is licensing per dongle, which allows the owner of the dongle to use the program on any computer. Licensing per server,

CPU, or points, regardless the number of users, is common practice as well as site or company licenses. Sometimes one can choose between perpetual (permanent) and annual license. For perpetual licenses one year of maintenance is often required, but maintenance (subscription) renewals are discounted. For annual licenses, there is no renewal; a new license must be purchased after expiration. Licensing can be host/client (or guest), mailbox, IP address, domain, and so on, depending on how the program is used. Additional users are inter alia licensed per extension pack (e.g., up to 99 users) which includes the base pack (e.g., five users). Some programs are modular, so one will have to buy a base product before they can use other modules.

Software licensing often also includes maintenance. This, usually with a term of one year, is either included or optional, but must often be bought with the software. The maintenance agreement (contract) contains minor updates (V.1.1 $\Rightarrow$ 1.2), sometimes major updates (V.1.2 $\Rightarrow$ 2.0) and is called update insurance or upgrade assurance. For a major update the customer has to buy an upgrade, if not included in the maintenance. For a maintenance renewal some manufacturers charge a reinstatement (reinstallment) fee retroactively per month, in case the current maintenance has expired. Maintenance normally doesn't include technical support. Here, one can differentiate between email and telephone support, also availability (e.g., 5 × 8, five days a week, eight hours a day) and reaction time (e.g., three hours) can play a role. This is commonly named gold, silver, and bronze support. Support is also licensed per incident as an incident pack (e.g., five support incidents per year).

Many manufacturers offer special conditions for schools and government agencies (EDU/GOV license). Migration from another product (crossgrade), even from a different manufacturer (competitive upgrade), is offered.

4.7.4 Free and Open-Source Software Licenses

There are several organizations in the FOSS domain who give out guidelines and definitions regarding software licenses. Free Software Foundation (FSF) maintains non-exhaustive lists of software licenses following their "The Free Software Definition" and licenses, which the FSF considers non-free for various reasons. The FSF additionally distinguishes between free software licenses that are compatible or incompatible with the FSF license of choice, the copyleft GNU general public license. The Open Source Initiative defines a list of certified open-source licenses following their "The Open Source Definition." Also the Debian Project has a list of licenses that follow their Debian Free Software Guidelines.

Free and open-source licenses are commonly classified into two categories: those with the aim to have minimal requirements about how the software can be redistributed (permissive licenses), and the protective share-alike (copyleft licenses).

An example of a copyleft free software license is the often used GNU general public license (GPL), also the first copyleft license. This license is aimed at giving and protecting all users unlimited freedom to use, study, and privately modify the software, and if the user adheres to the terms and conditions of the GPL, freedom to redistribute the software or any modifications to it. For instance, any modifications made and redistributed by the end user must include the source code for these, and the license of any derivative work must not put any additional restrictions beyond what the GPL allows.

Examples of permissive free software licenses are the BSD license and the MIT license, which give unlimited permission to use, study, and privately modify the software, and includes only minimal requirements on redistribution. This gives a user the permission to take the code and use it as part of closed-source software or software released under a proprietary software license.

It was under debate for some time if public domain software and public domain-like licenses can be considered a kind of FOSS license. Around 2004 lawyer Lawrence Rosen argued in the essay "Why the Public Domain Isn't a License" software could not truly be waived into public domain and can't therefore be interpreted as very permissive FOSS license, a position which faced opposition by Daniel J. Bernstein and others. In 2012 the dispute was finally resolved when Rosen accepted the CC0 as an open source license, while admitting that, contrary to his previous claims, copyright can be waived away, backed by Ninth Circuit decisions.

EXERCISES

1. Describe the *intellectual property issues (IPR)*.

2. Write a short note on the *Copyright Act and the Patent Law*.

3. Explain the semiconductor layout and design in detail.

4. Explain briefly the *IT Act of 2000* and the coverage in each chapter.

5. Briefly explain the ISO standards to be considered while laying down the policies.

6. Discuss patents, patentable inventions and non-patentable inventions.

7. Why do we require a policy for information security?

8. What are the ISO standards for information security? Briefly discuss.

9. Define the term "ISO." What are its security standards?

10. Explain the important features of WWW policy.

11. Comment on the importance of security policies in corporate environments.

12. State the benefits of email security policy.

13. Describe in detail the process of policy review.

14. What are the judicial aspects of cyber laws?

15. Write short notes on the following:
    (a) IT ACT of 2000
    (b) Patent Act
    (c) Copyright law

16. Explain the salient features of the SICLD Act.

17. Define the term "software licensing."

18. Provide an explanation for the term "ISO standards."

19. How are rights of consumers protected by the IT and Patent Acts?

20. What is cyber security? Explain some of the measures taken to ensure cyber security.

21. Justify the legal dimensions of information security.

CHAPTER 5

SECURITY OF EMERGING TECHNOLOGY[1]

This chapter will covert the Security of recent emerging technologies include Big Data Analytics, Cloud Computing, Internet of Things (IoT), Smart Grid, Supervisory control and data acquisition (SCADA) Control Systems, and Wireless Sensor Network (WSN)

5.1 SECURITY OF BIG DATA ANALYTICS

The *Big Data* is a large-scale information management and analysis technologies that exceed the capability of traditional data processing technologies. Big data can be defined by two or more characteristics of the following:

- **Volume:** A system is gathering large amounts of data
- **Variety:** The data being gathered and analyzed varies in structure and format
- **Velocity:** Data is gathered at a high speed
- **Value:** Significant value is derived from the analysis of data
- **Visibility:** Data is accessed or visible from disparate or multiple geographic regions
- **Variability:** Data flows can be highly inconsistent with periodic peaks

---

1   This chapter was published as Chapter 14 in Network Security and Cryptography. Sarhan Musa. Copyright ©2018 by Mercury Learning and Information LLC. All rights reserved.

- **Complexity:** Complexity of data when it is coming from multiple sources. The data must be linked, matched, cleansed, and transformed into required formats before actual processing.

Big Data can be Divided into Two Groups of Processing

- *Batch processing*: the analytics on data at rest (Hadoop for data volumes of desk)

- *Stream processing*: the analytics on data in motion (Storm for data volumes of memory)

Big Data analytics is the process of analyzing and mining Big Data. It can produce operational and products knowledge at an unprecedented scale and specificity. The technological advances in storage, processing, and analysis of Big Data can include the following:

a. Rapidly decreasing cost of storage and CPU power in recent years;

b. Flexibility and cost-effectiveness of datacenters and cloud computing for elastic computation and storage;

c. Development of new frameworks such as *Hadoop*, which allow users to take advantage of these, distributed computing systems storing large quantities of data through flexible parallel processing.

5.1.1 Big Data Analysis can Transform Security Analytics in the Following Ways

a. Accumulate data from various internal organizational sources as well as external sources to make a consolidated view of the required data into something called as a vulnerability database.

b. Perform an in-depth analytics on the data using security intelligence hence uncovering unique patterns that could be the source of many security issues.

c. One dimensional view of all the related information.

d. Real time analysis of streaming data and uses previous results as a feedback to the system as a whole.

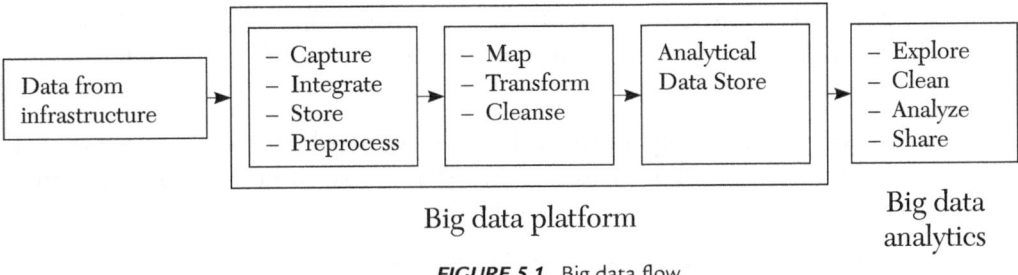

FIGURE 5.1. Big data flow.

5.1.2 Big Data Analytics for Security Issues and Privacy Challenges

- **Protected database storage and transaction log file:** Availability and scalability have required auto tiring for the big data management. Auto tiring solution do not keep track of where the database is actually stored, which act as new demanding to protected database storage.

- **Secure computations in distributed frameworks:** Parallelism is used in computations and physical storage to process very large data. MapReduce framework is an example. Protecting the mappers and protecting the data in the presence of untrusted mapper are two major attack prevention measures.

- **Privacy issues for non-relational data stores:** No SQL database embedded protection in the middleware. It does not provide any type of support for enforcing it explicitly in the database. However, gathering aspect of No SQL databases impose additional demanding to the strength of such privacy practice.

- **End-point input validation/filtering:** This method is used to identify the trusted data and to verify that source of data input details is not despiteful.

- **Extensible and pasture able privacy preserving data mining and analysis:** This method is used to troubling manifestation by the possibly enabling appropriation of security, forward marketing; reduce civil freedoms, and increase state and corporate control. Anonymzing data for analysis is not sufficient to manage user security.

- **Real-time security and compliance monitoring:** This method gives the number of alerts generated by privacy devices. These alerts lead to many false positives, which are mostly ignored. It is used to provide real-time problem detection based on scalable privacy analysis.

- **Granular audits:** This method is used for compliance, regulation and forensics reasons. It is used to deal with the data object, which probably are allocated.

- **Cryptographically enforced access control and secure communication:** This method is used to ensure fairness, authentication, and agreement among the distributed entities.

- **Granulated access control:** Secrecy-secures access to data by people that should not have access. Granulated access control give data manager a skiver in place of a blade to share data as much as possible without agree privacy.

- **Information security:** How to tackle big data from security point of view is a hard task. Thus, information security is one of the big data issues.

- **Metadata provenance:** Data will increase in complexity due to large provenance graphs generated from provenance enabled programming environments in big data applications. Analysis of such large provenance graphs to detect metadata dependencies for security and confidentiality purpose is computationally intensive.

5.2 SECURITY OF CLOUD COMPUTING

- Cloud Computing is a type of computing in which services are delivered through the Internet which depends on sharing of computing resources than having local servers or personal devices to handle the applications.

- Cloud Computing make use of increasing computing power to execute millions of instructions per second.

- Cloud Computing uses networks of a large group of servers with specialized connections to distribute data processing among the servers. Instead of installing a software suite for each computer, this technology requires to install single software in each computer that allows users to log into a Web-based service and which also hosts all the programs required by the user.

- Local computers no longer have to take the entire burden when it comes to running applications.

- Cloud computing technology is being used to minimize the usage cost of computing resources.

- The cloud network, consisting of a network of computers, handles the load instead. The cost of software and hardware on the user end decreases. The only thing that must be done at the user's end is to run the cloud interface software to connect to the cloud.

Cloud Computing Consists of Two Ends

- **Front end:** It includes the user's computer and software required to access the cloud network

- **Back end:** It consists of various computers, servers and database systems that create the cloud.

The user can access applications in the cloud network from anywhere by connecting to the cloud using the Internet. Some of the real time applications which use Cloud Computing are Gmail, Google Calendar, and Dropbox etc.,

5.2.1 Cloud Deployment Models

- **Public Cloud:** The cloud infrastructure is typically owned by an organization selling cloud services known as a cloud service provider (CSP), and delivered to the general public on a subscription basis.

- **Private Cloud:** The cloud infrastructure is operated only for an individual organization. The infrastructure may be managed by the organization itself or by a third party, and it may be located either on-premises or off.

- **Community Cloud:** The cloud infrastructure is shared by several organizations and supports a specific community that has shared concerns. The infrastructure may be managed by the organizations or by a third party and may be located on-premise or off-premise.

- **Hybrid Cloud:** The cloud infrastructure is a combination of both private and public cloud instances that remain unique entities, but are bound together by standardized or proprietary technology that enables data and application portability.

5.2.2 The Three Layers of Cloud Computing Services Model (Software, Platform or Infrastructure (SPI) Model)

- **Software as a Service (SaaS):** provides a way of delivering centrally hosted applications over the Internet as a service. The user consumes a software application across the Internet. The user has no infrastructure or applications to manage and update, no setup or hardware costs, and application accessibility from any Internet connection. It is a user interfaces. It is Network-hosted application. For example, IBM Lotus Live.

- **Platform as a Service (PaaS):** provides the service and management that is similar to the operating system. The CSP provides an additional layer on top of the infrastructure. Services include operating system, network access, storage, database management systems, hosting, server-side scripting, and support. The user can use this environment and the tools provided to create software applications. It is Network-hosted software development platform. For example, Google App Engine (GAE) and Microsoft Azure.

- **Infrastructure as a Service (IaaS):** provides the cloud services of the basic hardware, such as CPU, Network, storage and so on. the CSP provides the virtualized computing infrastructure. This generally includes virtual compute instances, network connectivity, IP infrastructure, bandwidth, load balancers, and firewalls. The user is responsible for installing and maintaining everything above the hypervisor (from the operating system upward). Provider hosts customer Virtual Machine (VMs) or provides network storage. For example, Amazon Web Service (AWS).

5.2.3 Security Concerns and Challenges of Cloud Computing

- **Authentication:** Applications and data are accessed over the internet which increases the complexity of the authentication procedures.

- **Authorization:** It is computing environment requires the use of the cloud service and providers' services for identifying the access policies.

- **Data Integrity:** Appropriate mechanics are requires for detecting accidental and intentional changes in the data.

- **Security of data while at rest:** Appropriate separation procedures are required to ensure the isolation between applications and data from different organizations.

- **Security of data while in motion:** Appropriate security procedures are required to ensure the security of data while in motion.

- **Auditing:** Appropriate auditing procedures are required to get visibility into the application, data accesses, and actions preformed by the application users.

5.2.4 Cloud Security as Consumers Service

- Identity Services and Access Management Services
- Data Loss Prevention (DLP)
- Web Security
- Email Security
- Security Assessments
- Intrusion Management, Detection, and Prevention (IDS/IPS)
- Security Information and Event Management (SIEM)
- Encryption
- Business Continuity and Disaster Recovery
- Network Security

5.3   SECURITY OF INTERNET OF THINGS (IoT)

- Internet of Things (IoT) is the concept of things, users and cloud services getting connected via the Internet to enable and provide intelligent services for users. Figure 5.2 shows the IoT concept topology.

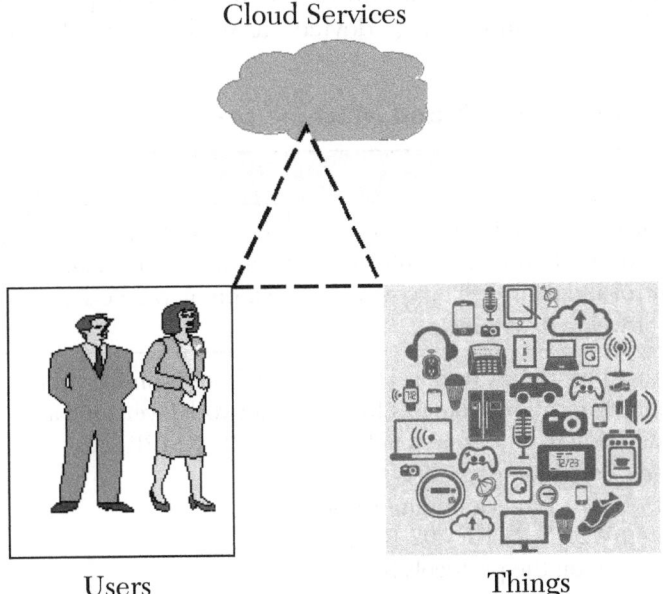

FIGURE 5.2. IoT concept topology.

5.3.1 Evolution of IoT

- Internet service providers (ISPs)
- Radio frequency identification (RFID)
- Application service providers (ASPs)
- Software as a Service (SaaS)

5.3.2 Building Blocks of the Internet of Things (IoT)

- **Sensors/Actuators:** Sensors and actuators are the tools that allow us to monitor and collect data, and control the THINGS in the IoT.

- **Devices:** Simply put, devices are the THINGS. Using sensors and actuators, these devices will be more intuitive and efficient than we ever thought possible.

- **Gateways:** IoT gateways will help devices intelligently communicate for greater efficiency, intuitive data management/classification, and increased security.

- **Master of Devices and Service Providers:** For every device or service in the IoT, there must be a master. This could be the device manufacturer, a cloud service provider, and or an IoT solution provider. The master's role is to issue and manage devices, as well as facilitate data analysis.

5.3.3 Different Between IoT and Machine-to-Machine (M2M)

M2M	IoT
It focused on connecting machines (or devices) for use in remote mentoring and control and data exchange–mainly proprietary closed systems	It is about corresponding the way humans and machines connect using common public services
Uses either proprietary or non-IP based communication protocols for communication within its area network such as ZigBee, Bluetooth, Z-Wave, ModBus, Power Line Communication (PLC), 6LoWPAN, IEEE 802.15.4, etc. It focuses on the protocols below the network layer	It focuses on the protocols above the network layer such as HTTP, CoAP, DDS, XMPP, etc.

(continued)

M2M	IoT
Data is collected in point solutions and can be accessed by on site applications and storage infrastructure	Data is collected in the cloud and can be accessed by cloud applications
Focused on hardware with embedded modules	Focused on software

5.3.4 IoT Layers Models

5.3.4.1 Three Layer Model

- **Application layer:** Information availability, user authentication, information privacy, data integrity, IoT platform stability, middleware security, management platform. For example, remote medical services, cloud computing, smart grid, smart traffic, smart home, environment monitoring

- **Transport layer:** DOS/DDOS attacks, forgery/middle attack, heterogeneous network attacks, WLAN application conflicts, capacity and connectivity issues, etc. For example, LAN, Ad hoc, GPRS, WiFi, 3G/4G

- **Sensing layer:** collect and process the data from the physical world. Interruption, interception, modification, fabrication, uniform coding for RFID, conflict collision for RFID, etc. For example, WSN, RFID, RSN, MEMS, GPS

5.3.4.2 Four Layer Model

- **Service layer:** provides the interface and communicates with the users. For example, health care, smart home, safety, smart industry. It uses light weight security for security.

- **Platform layer:** supports the IoT applications and services. For example, interface, context awareness, operating system, cloud services. It uses privacy preserving for security.

- **Network layer:** serves to transmit the data among devices, contents, services and users. It processes, controls, and manages enormous amounts of network traffic. For example, Context connection, context based network, group mobility, gateway PnP. It uses Authentication for security.

- **Device layer:** perceives the environment with various sensing devices, processes it to send to the sink node or gateway, and responds to it if necessary. For example, actuator, sensor, resource management, automatic control. It uses sensor data integrity for security.

5.3.4.3 Seven Layer Model

- **Physical Devices and Controllers (Layer 1):** It controls multiple devices. These are the "things" in the IoT, and they include a wide range of endpoint devices (computing nodes) that send and receive information, e.g., smart controllers, sensors, RFID readers, etc., and different versions of RFID tags. Data confidentiality and integrity must be taken into account from this level upwards. Secure content (silicon)

- **Connectivity (Layer 2):** Communication and processing unit. Reliable and timely information transmission. Secure network access (hardware & protocols)

- **Edge (Fog) Computing (Layer 3):** Data element analysis and transformation. Secure communications (protocols and encryption)

- **Data Accumulation (Layer 4):** Storage and making network data usable by applications. Tamper resistant (software)

- **Data Abstraction (Layer 5):** Aggregation and Access. Also, abstracting the data interface for applications. Secure storage (hardware & software)

- **Application (Layer 6):** Reporting, analytics, and control. Authentication/Authorization (software)

- **Collaboration and Processes (Layer 7):** Involving people and business processes. Identity Management (software)

5.3.5 Applications of IoT

1. **Wearable**
   - Entertainment
   - Fitness
   - Smart watch
   - Location and tracking

2. **Health Care**
   - Remote monitoring
   - Ambulance telemetry
   - Drugs tracking

- Hospital asset tracking
- Access control
- Predictive maintenance

3. **Building & Home Automation**
   - Access control
   - Light and temperature control
   - Energy optimization
   - Predictive maintenance
   - Connected appliances

4. **Smart Cities**
   - Residential E-meters
   - Smart street lights
   - Pipeline leak detection
   - Traffic control
   - Surveillance cameras
   - Centralized and integrated system control

5. **Automotive**
   - Infotainment
   - Wire replacement
   - Telemetry
   - Predictive maintenance
   - Car to Car (C2C) and Car to Infrastructure (C2I)

6. **Smart Manufacturing**
   - Flow optimization
   - Real time inventory
   - Asset tracking
   - Employee safety
   - Predictive maintenance
   - Firmware updates

Every connected device creates opportunities for attackers. These vulnerabilities are broad, even for a single small device. The risks posed include data transfer, device access, malfunctioning devices, and always-on/always-connected devices. The main challenges in security remain the security limitations associated with producing low cost devices, and the growing number of devices which creates more opportunities for attacks.

5.3.6 New Challenges Created by the IoT

- **Security:** Prevent attackers and hacker from accessing the IoT

- **Privacy:** protect identity and privacy data from attackers and hacker access IoT

- **Interoperability and standards:** ensure that devices communicate securely by IoT manufacturers and ASP developers

- **Legal and regulatory compliance:** contribute toward legal, tax, and regulatory requirements regarding IoT-related business transactions that involve payment for goods and services by the international, federal, and state levels.

- **E-commerce and economic development issues:** connectivity and information sharing to be deployed globally by IoT and the economic rules of engagement for conducting business on the World Wide Web.

5.3.7 Security Requirements of IoT

- **Confidentiality:** Ensuring that only authorized users access the data and information

- **Integrity:** Ensuring completeness, accuracy, and absence of unauthorized data manipulation

- **Availability:** Ensuring that all system services are available, when requested by an authorized user

- **Accountability:** An ability of a system to hold users responsible for their actions and operations

- **Auditability:** An ability of a system to conduct persistent monitoring of all actions and operations

- **Trustworthiness:** An ability of a system to verify an accurate identity and establish trust in a third party

- **Non-repudiation:** An ability of a system to confirm occurrence/non-occurrence of an action and an operation

- **Privacy:** Ensuring that the system obeys privacy policies and enabling individuals to control their personal data and information

5.3.8 Three Primary Targets of Attack against IoT

- **Attacks against a device:**

  An attacker takes advantages of IoT devices because many of the devices will have an inherent value by the simple nature of their function.

  As devices will be trusted with the ability to control and manage things, they also present a value for their ability to impact things. The devices have a value based on what is entrusted to those devices.

- **Attacks against the communication between devices and masters:**

  An attack involves monitoring and altering messages as they are communicated. The volume and sensitivity of data traversing the IoT environment makes these types of attacks especially dangerous, as messages and data could be intercepted, captured, or manipulated while in transit. All of these threats endanger the trust in the information and data being transmitted, and the ultimate confidence in the overall infrastructure.

- **Attacks against the masters:**

  Attacks against manufacturers, cloud service providers, and IoT solution providers have the potential to inflict the most amount of harm. These masters will be entrusted with large amounts of data, some of it highly sensitive in nature. This data also has value to the IoT providers because of the analytics, which represent a core, strategic business asset—and a significant competitive vulnerability if exposed. Disrupting services to devices also poses a threat as many of the devices will depend on the ability to communicate with the masters in order to function. Attacking a master also presents the opportunity to manipulate many devices at once, some of which may already be deployed in the field.

5.3.9 Hybrid Encryption Technique

Hybrid encryption technique is for information integrity, confidentiality, and being non-repudiation in data exchange for IOT.

A. Creating a Key

- Key production process in AES is used to create a key

- Two four by four matrixes (stay and key) are used to produce key for encryption

- Choose a place from the state matrix and a key from the key matrix randomly and produce public key of H by sender in XOR operation

- Produced key of h is on the basis of hexadecimal. Then public key h is produced.

B. Encryption

- A message sent from the sender to the receiver is in a multinomial called message. After making a multinomial message, the sender randomly choose a multi nominal like r from the collection like L_r.

- We can have a message by multi nominal r. therefore, it should not be revealed by the sender.

$$\text{Encryption} = p_r \times h + \text{message} \tag{5.1}$$

This message will be transmitted to the receiver as an encryption message with security capability.

C. Decryption

- When the message is encrypted, the receiver tries to open the message by its private key or encrypt the message.

- The receiver has both private keys: f and f_p. In fact f_p is conversed with multinomial of f, so it can be concluded that it will be $f \times f_p = 1$ message receiver multiplies a message on the part of private key that is displayed below with the parameter a:

$$a = f \times \text{encryption} \tag{5.2}$$

$$a = f \times (p_r \times h + \text{message}) \tag{5.3}$$

$$a = f \times P_r \times h + f \times \text{message} \tag{5.4}$$

- To choose a correct parameter, coefficients of the poly nominal formula between $-q/2$ and $p/2$ are selected.

- As p = 3, then it drastically reduce and does not have any effect on the process:

$$p_r \times h = 0 \tag{5.5}$$

$$a = f \times \text{message} \tag{5.6}$$

- Parameter b will be calculated. Just multiply private key f in initial message which has been sent by the sender:

$$a = b = f \times message \tag{5.7}$$

$$Decryption = (f_p \times b) / x^2 \tag{5.8}$$

- Whenever Decryption = Message, we will be sure that the message will reach security to the recipient without any disorder

D. Digital signature

- Digital signature is used for message validity and proof of identity and security.

- We must go from the sender to the receiver, so the receiver of the former step acts as the sender now and the sender of the former step acts as the receiver.

$$Encryption \; sign = (message * f)/x^2 \tag{5.9}$$

$$Decryption = (h / 2 \times f_p \times Encryption \; sign) /2 \times h \tag{5.10}$$

5.3.10 Hybrid Encryption Algorithm Based on DES and DSA

- Regroup 64-bit data according to blocks and put the output into L_0, R_0 two parts, each of 32 bits, and its replacement rule is that exchange the 58th bit with the first, the 50th bit with the second, ..., and so on, the last one is the original No. 7.

- L_0 and R_0 are the two parts after the transposition output, L_0 is the left 32-bit of the output, and R_0 is the right 32 bits.

- Sub-key generation algorithm: The 8th, 16th ...... 64th bit is a parity bit according to DES algorithm rule, and is not involved in DES operation. Key actual use 56 bits, this 56 is Divided into two parts C_0 and D_0, each of 28 bits, and cycle left for the first time, to obtain C_1 and D_1, then the C_1 (28 bits), D_1 (28 bits) obtained were combined to form a 56-bit date, and then selection transposition 2 through the narrow, so as to get a key K_0 (48 bits). And so on, K_1, K_2... K_{15} can be got.

- Calculated date hash value of the encrypted cipher text using the secure hash algorithm SHA-1. Parameters used in DSA signature are:
  - p: primes of L bits long. L is a multiple of 64 bits; the range is 512~1024 bits;
  - q: prime factors of 160bits of p − 1;

- g: $g = h^{((p-1)/q)} \bmod p$, h satisfy $h < p - 1$, $h^{((p-1)/q)} \bmod p > 1$;
- x: $x < q$, x is the private key;
- y: $y = g^x \bmod p$, (p, q, g, y) are public keys;

Signature process is:

P: generates a random number k, $k < q$;
P: compute $r = (g^k \bmod p) \bmod q$ and $s = (k^{(-1)}(H(m) + xr)) \bmod q$
The result of the signature is (m, r, s).

Where H (m) is the Hash value of m, m is plaintext to be signed or Hash value of the plaintext. The final signature is integers (r, s), which is sent to the authenticator.

- Calculate the encrypted digital signature use SHA-1 algorithm after the reader receives the cipher text, if the signatures consistent with that provided by the sender, then use decrypt the cipher text with the sub-key generated by the DES algorithm to generate plaintext.

5.3.11 Advance Encryption Standard (AES)

AES is based on a design principle known as a substitution-permutation network, combination of both substitution and permutation, and is fast in both software and hardware. Unlike its predecessor DES, AES does not use a Feistel network. AES uses Rijndael cipher which has a fixed block size of 128 bits, and a key size of 128, 192, or 256 bits and is defined in three versions 10, 12, and 14 rounds respectively.

There are three steps performed in AES
- encryption,
- decryption
- key generation.

AES Encryption

Step 1: Get the plain text and the key
Step 2: Perform the pre-round transformation using the plain text
Step 3: With "n" key length, perform transformation for "n" rounds
Step 4: cipher text achieved

AES Decryption

Repeat the step followed in encryption in reverse order:

Step 1: cipher text achieved
Step 2: With "n" key length, perform transformation for "n" rounds
Step 3: Perform the pre-round transformation using the plain text
Step 4: Get the plain text and the key

Key Generation

Step 1: Get the key
Step 2: based upon number of round, calculate required number of words
Step 3: In an array of 4 bytes, first four words are made from the key
Step 4: Get the next word
Step 5: Repeat step 4 until required number of words are reached.

5.3.12 Requirements for Lightweight Cryptography

- **Size (circuit size, ROM/RAM sizes):** determinates the possibility of implementation in a device.

- **Power:** especially important with the RFID and energy harvesting devices

- **Power consumption:** It is important with battery-driven devices.

- **Processing speed (throughput, delay):** A high throughput is necessary for devices with large data transmissions, while a low delay is important for the real-time control processing.

5.3.13 Lightweight Cryptography in the IoT

- **Efficiency of end-to-end communication:** Application of the lightweight symmetric key algorithm allows lower energy consumption for end devices.

- **Applicability to lower resource devices:** The lightweight cryptographic primitives would open possibilities of more network connections with lower resource devices.

5.3.14 Prevention of Attacks on IoT

Cyber attackers are exploiting vulnerabilities on IoT devices in an increasing number of distributed denials of service (DDoS) attacks. The Prevention of attacks on IoT can summarized five steps as below:

- **Changing the default credentials of IoT devices:** Devices with weak or default credentials are vulnerable to compromise. IoT devices should be secured with strong authentication to avoid brute force attacks.

- **Disabling universal plug and play (UPnP) on gateway routers:** UPnP allows ports inside a network to be opened easily. Using UPnP, external computers are able to communicate to devices inside the network. To prevent this, we should disable UPnP on gateway routers. Some applications may be affected by disabling UPnP, so reconfiguration could be necessary.

- **Update IoT devices frequently:** Update IoT devices with the latest firmware and patches as soon as possible to ensure that the known vulnerabilities are addressed.

- **Ensure proper firewall configuration and identify malicious traffic:** Configure the firewall to block incoming User Datagram Protocol (UDP) packets because they are used to exploit IoT devices.

- **Review reliance on easily identified internet connections:** Examine the level of reliance on public-facing web servers that are easy to identify externally and that are used for critical operations. It is important to review incident response procedures as well so that operations are not halted due to a cyber attack. In addition, IoT devices that are on public-facing servers should be secured to prevent unauthorized access.

5.4 SECURITY OF SMART GRID

- Smart grids utilize communication technology and information to optimally transmit and distribute electricity from suppliers to consumers. It is the next generation power system.

- The grid environment requires security mechanisms that have the following characteristics:
  - cross multiple administrative domains,
  - have high scalability in terms of a large and dynamic user population,
  - support a large and dynamic pool of resources each with probably different authentication and authorization policies,
  - have the ability of grid applications to acquire and release resources dynamically during execution

5.4.1 Smart Grid Challenges

- the network congestion and safety related factors;
- the lack of pervasive and effective communications, monitoring, fault diagnostics, and automation;
- power grid integration, system stability, energy storage, which are introduced by the adaptation of renewable and alternative energy sources.

5.4.2 Smart Grid Layers

- Master station system Layer
- Remote communication network Layer
- Terminal Layer
- Cross (Life cycle of information systems) Layer
- Security management Layer

5.4.3 Information Security Risks and Demands of Smart Grid

- **Master station system Layer:**
  - Physical layer attacks and protection
  - Network layer attacks and protection
  - Host layer attacks and protection
  - Application layer attacks and protection
  - Data leaks and prevention; Backup and recovery
  - Cloud computing application and its Risks
  - Intercept and anti-intercept

- **Remote communication network Layer:**
  - Monitor and anti-monitor
  - Tamper and anti-tamper
  - Encrypted communication channel
  - Fake terminal
  - Fake master station
  - Terminal integrity
  - Terminal network security

- **Terminal Layer:**
  - Lack of computing, storage, and process resources to implement security schemes

- Internet of Things application and its risks
- Lack of security consideration in system planning and system analysis stages
- Lack of security design in system design stage

- **Cross (Life cycle of information systems) Layer:**
  - Code security and secure system development in system implementation stage
  - Security management in system running and maintenance stage
  - Sensitive information processing in system obsolescence stage
  - Social engineering attacks
  - How to build an effective Information

- **Security management Layer:**
  - Security Management Systems for smart grid

5.4.4 Smart Grid Security Objectives

- **Availability:** Ensuring timely and reliable access to and use of information is of the most importance in the Smart Grid.

- **Integrity:** Guarding against improper information modification or destruction is to ensure information non-repudiation and authenticity.

- **Confidentiality:** Preserving authorized restrictions on information access and disclosure is mainly to protect personal privacy and proprietary information.

5.4.5 The Smart Grid System can be Divided in Three Major Systems

- Smart Infrastructure System,
- Smart Management System,
- Smart Protection System.

These major systems are also subdivided in other subsystems, applications and/or objectives, as show in Figure 5.3.

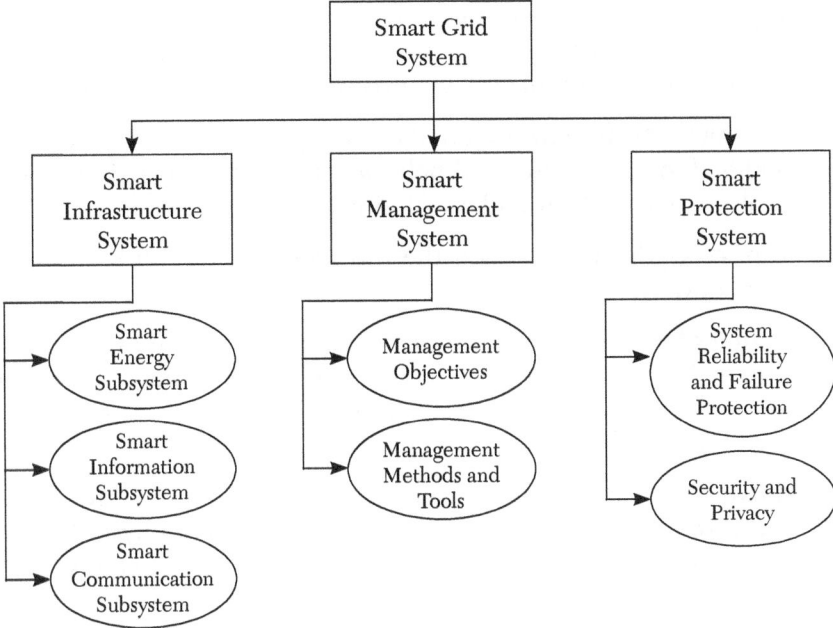

FIGURE 5.3. Smart grid system classifications.

5.4.6 Types of Security Attacks that can Compromise the Smart Grid Security

- **Passive Attacks:** They aim to learn and use the system information without affecting the system resources. The attack target is only the transmitted information in order to learn the system configuration, architecture, and normal operation behavior.

- **Active Attacks:** They are planned to affect the system operation through data modification or introducing false information into the system.

5.4.7 Cybersecurity Attacks in a Smart Grid

- **Eavesdropping:** This is a passive attack described as an unauthorized interception of an on-going communication without the consent of the communication parties.

- **Traffic Analysis:** Similar to eavesdropping attack, but the attacker monitors the traffic patterns in order to infer useful information from it.

- **Replay:** This attack consists of capture transmitted messages and their retransmission in order to cause an unauthorized effect. The retransmitted messages are normally valid except the timestamp field.

- **Message Modification:** Similar to replay attack but the message is modified to cause unwanted behavior in the system. This attack can also involve message delay and reordering a message stream.

- **Impersonation:** It is when the intruder pretends to be an authorized entity or device.

- **Denial of Service:** It aims to suspend or interrupt the system communications. To accomplish this effect, the attacker can flood the communication network with messages to disable the physical components access, inhibiting the system's normal operation.

- **Malware:** It aims to exploit internal weaknesses of the system with the goal of steal, modify, and destroy information and/or physical components of the system. It can also obtain unauthorized access to the system.

5.5   SECURITY OF SCADA CONTROL SYSTEMS

- A Supervisory Control and Data Acquisition (SCADA) system is a common process automation system which is used to gather data from sensors and instruments located at remote sites and to transmit data at a central site for either control or monitoring purposes.

- SCADA systems evolved from hardware and software that include standard PCs and operating systems, TCP/IP communications, and Internet access. SCADA system can monitor and control hundreds to hundreds of thousands of I/O points.

- SCADA systems differ from Distributed Control Systems (DCSs), DCSs cover plant sites. While SCADA systems cover much larger geographic areas.

- SCADA architecture supports TCP/IP, UDP or other IP-based communications protocols as well as strictly industrial protocols such as Modbus TCP, Modbus over TCP or Modbus over UDP, all working over private radio, cellular or satellite networks.

5.5.1 Components of SCADA Systems

- Instruments that sense process variables

- Operating equipment connected to instruments

- Local processors that collect data and communicate with the site's instruments and operating equipment Programmable Logic Controller (PLC), Remote Terminal Unit (RTU), Intelligent Electronic Device (IED), or Process Automation Controller (PAC)

- Short range communications between local processors, instruments, and operating equipment

- Host computers as central point of human monitoring and control of the processes, storing databases, and display of statistical control charts, and reports. Host computers are also known as Master Terminal Unit (MTU), the SCADA server, or a PC with Human Machine Interface (HMI)

- Long range communications between local processors and host computers using wired and/or wireless network connections.

5.5.2 SCADA System Layers

- **Supervisory Control Layer:** The supervisory control layer (the control center) is responsible for monitoring the operation of SCADA systems by gathering data from field devices, performing control and supervisory tasks, and sending control commands to field controllers through the communication network. The supervisory control layer of SCADA system consists of following elements:

  - SCADA server
  - builder server
  - communication server
  - database server
  - diagnostic server
  - application server
  - human–machine interface
  - system operators

- **Automatic Control Layer:** The automatic control layer (regulatory control layer) is in charge of regulating the operation of physical processes based on control commands from the control center and sensor measurements from field devices. The control signals are then transmitted to field devices through the communication network. Various system variables, including control commands, sensor measurements, and control signals, are gathered within the control center for supervisory and management purposes. The automatic control layer of SCADA system consists of following elements:

- – master terminal units (MTUs)
- – remote terminal units (RTUs)
- – programmable logic controllers (PLCs)
- – intelligent electronic devices

- **Physical Layer:** The physical processes (e.g., electric power grids, gas pipelines, and water networks) are equipped with actuators (e.g., motors, compressors, pumps, and valves), sensors (e.g., temperature sensors, pressure sensors, flow sensors, level sensors, and speed sensors), and protection devices (e.g., circuit breakers and protective relays) to realize technological goals. The physical elements are controlled and monitored by the control center through the automatic control layer and the communication network.

5.5.3 Requirements and Features for the Security of Control Systems

- Critical path protection

- Strong safety policies and procedures

- Knowledge management

- System development skills

- Enhanced security for device

- Sensor networks solutions

- Operating system based on microkernel architecture

- Increasing quality of software with security features

- Security requirements early in the software development cycle

- Compliance to standards for software development

- Integration of different technologies

- Vulnerability analysis based on proactive, discovery, and adaptation solutions

- Innovative risk management approaches

- Ensure authentication, confidentiality, integrity, availability, and non-repudiation

- Calculate risk as impact to security and safety

5.5.4 Categories for Security Threats to Modern SCADA Systems

- Insiders
- Hackers
- Hidden criminal groups
- Nation-states

5.6 SECURITY OF WIRELESS SENSOR NETWORKS (WSNs)

- A Wireless Sensor Networks (WSN) is an infrastructure-less network composed of large number of sensor nodes. These cooperatively sense and control the environment to enable its interaction with people or devices.

- Wireless Sensor Networks (WSNs) are considered as one of the core technologies in implementing Cyber Physical Systems (CPSs).

- WSNs include sensor nodes, actuator nodes, gateways and clients. A large number of sensor nodes deployed randomly inside of or near the monitoring area, form networks through self-organization.

- The data is captured at the level of the sensor node, compressed and transmitted to the gateway. Through the gateway connection, data is then passed by the base station to a server.

- Sensor nodes monitor the collected data to transmit along to other sensor nodes by hopping. During the process of transmission, the monitored data may be handled by multiple nodes to get to gateway node after multi-hop routing, and finally reach the management node through the internet or satellite.

- It is the user who configures and manages the WSN with the management node, publish monitoring missions, and collection of the monitored data.

5.6.1 WSN Layers

- **Transport Layer:** Responsible for managing end-to-end connections. Reliable transport of data.

- **Network and Routing Layer:** Responsible for routing of sensors based on addressing and location awareness, sensor networking, power efficiency, and topology management. It provides more effective routing the data From "Node to node, node to sink, node to base station and node to Cluster head & vice versa." Due to the broadcast method every node works as a router.

- **Data Link Layer:** Responsible for the multiplexing of data streams, data frame detection, medium access, and error control

- **Physical Layer:** Responsible for frequency selection, carrier frequency generation, signal detection, modulation, and data encryption

5.6.2 Security Requirements in WSNs

- **Data Confidentiality:** To provide the data confidentiality, an encrypted data is used so that only recipient decrypts the data to its original form. Only the authorized sensor nodes can get the content of the messages.

- **Data Integrity:** Data received by the receiver should not be altered or modified. Original data is changed by intruder or due to harsh environment. The intruder may change the data according to its need and sends this new data to the receiver. Thus, any received message has not been modified in send by unauthorized parties.

- **Data Authentication:** It is the procedure of confirmation that the communicating node is the one that it claims to be. The receiver node needs to make verification that the data is received from an authenticate node.

- **Data Availability:** The services are available all the time whenever necessary.

- **Source Localization:** For data transmission some applications use location information of the sink node. It is essential to give security to the location information. Non secured data can be controlled by the malicious node.

- **Self-Organization:** In WSN no fixed infrastructure exists, hence, every node is independent having properties of adaptation to the different situations and maintains self organizing and self healing properties.

- **Data Freshness:** Each message transmitted over the channel is new and fresh. It guarantees that the old messages cannot be replayed by any node.

This can be solved by adding some time related counter to check the freshness of the data.

5.6.3 The Attacks Categories in WSNs

- **Outsider Vs Insider Attacks:** Outsider attacks are external attacks and the insider attacks are internal attacks. An outsider attacks come from outside the WSN. With the help of Outsider attack the bad data is inserted in the network for the services interruption. An insider attack is also known as the internal attack, these attacks come from the inside of the WSN.

- **Passive Vs Active Attacks:** Passive attacks include eavesdropping on or monitoring packets exchanged within the WSN; in active attacks, an attacker has the capability to remove or modify the messages during the transmission on the network.

- **Mote-class Vs Laptop-class Attacks:** In Monte-class attacks, an adversary attacks a WSN by utilizing a few nodes with similar capabilities to the network nodes; in laptop-class attacks, an adversary can use more powerful devices to attack a WSN. These devices have greater transmission range, processing power, and energy reserves than the network nodes.

5.6.4 Attacks and Defense in WSNs at Different Layers

Layer	Possible Type of Attacks	Defense
Transport	Flooding	Client puzzles
	Desynchronization	Authentication
Network and Routing	Spoofed, altered or replayed routing information	Egress filtering, authentication, monitoring
	Selective forwarding	Redundancy, probing
	Sinkhole	Authentication, monitoring, redundancy
	Sybil	Authentication, probing
	Wormholes	Authentication, packet leashes by using geographic and temporal information
	Hello flood attacks	Authentication, verify the bidirectional link
	Acknowledgment spoofing	Authentication

(continued)

(continued)

Layer	Possible Type of Attacks	Defense
Data Link	Collision	Error-correcting code
	Exhaustion	Rate limitation
	Unfairness	Small frames
Physical	Jamming	Spread-spectrum, priority messages, lower duty cycle, region mapping, mode change
	Tampering	Tamper-proofing, hiding

5.6.5 Security Protocols in WSNs

- **Sensor Protocols for Information via Negotiation (SPINs):**

    – SPIN is an adaptive routing protocol, which transmits the information first by negotiating.

    – The SPIN transmission is data centric; it is only transmitted to the nodes that have interest in the data. This process continues until the data reaches the sink node. SPIN reduces both the network overhead and the energy consumption in the transmission. There will not be duplicate messages in the network since nodes negotiate before transmitting the data.

    – SPIN make use of metadata of the actual data to be sent. Metadata will contain the description of the message that the node wants to send. The actual data will be transmitted only if the node wishes to receive it. SPIN makes use of 3 messages namely,

        1. ADV: Before sending a message, a node first generates the descriptor of the message to be sent. This metadata is exchanged by making use of ADV message. ADV message informs the size, contents and requirements of the message. This helps the receiving node on deciding transmission of the message.

        2. REQUEST: After receiving the ADV message receiver node verifies the descriptor whether the message is a duplicate and whether receiver node's battery capabilities are enough to transmit the data. If the node is interested in data, it replies with a REQUEST message to the sender node.

3. DATA: If the sender node receives a REQUEST message, it starts the actual transmission of data by making use of DATA message. This is the actual data transfer phase.

- **Localized Encryption and Authentication Protocol (LEAP)**

  - LEP is a protocol with key management scheme that is very efficient with its security mechanisms used for large scale distributed sensor networks. It is designed to support for in network processing such as data aggregation. In-network processing results in reduction of the energy consumption in network. To provide the confidentiality and authentication to the data packet, LEAP uses multiple keys mechanism. For each node four symmetric keys are used as follows:

    Individual Key: Used for the communication between source node and the sink node.

    Pair wise Key: Shared with another sensor nodes.

    Cluster Key: Used for locally broadcast messages and shares it between the node and all its surrounding neighboring nodes.

    Group Key: Used by the entire network Nodes.

- **TINYSEC**

  - TINYSEC is a lightweight protocol and link layer security architecture for WSNs.

  - It supports integrity, confidentiality and authentication. To achieve confidentiality, encryption is done by using CBC (Cipher-block chaining) mode with cipher text stealing, and authentication is done using CBC-MAC. No counters are used in TINYSEC. Hence, it doesn't check the data freshness. Authorized senders and receivers share a secret key to compute a MAC.

  - TINYSEC has two different security options. One is for authenticated and encrypted messages (TinySec-AE) and another is for authenticated messages (TinySec-Auth).

  - In TinySec-AE, the data payload is encrypted and the received data packet is authenticated with a MAC.

  - In TinySec-Auth mode, the entire packet is authenticated with a MAC, but on the other hand the data payload is not encrypted.

- In CBC, Initialization Vector (IV) is used to achieve semantic security. Some of the messages are same with only little variation. In that case IV adds the variation to the encrypted process. To decrypt the message receiver must use the IV. IVs are not secret and are included in the same packet with the encrypted data.

- **ZIGBEE**

  - ZIGBEE is a worldwide open standard for wireless radio networks in the monitoring and control fields.

  - IEEE 802.15.4 is a standard used for ZIGBEE. The IEEE 802.15.4 standard defines the characteristics of the physical and MAC layers for Low-Rate Wireless Personal Area Networks (LR-WPAN).

  - To implement the security mechanism ZIGBEE uses 128 bit keys. A trust center is used in ZIGBEE which authenticates and allows other devices/nodes to join the network and also distribute the keys.

  - Three different roles in ZIGBEE are:

    Trust Manager: It authenticates the devices which are requesting to join the network.

    Network Manager: It manages the network keys and helps to maintain and distribute the network keys.

    Configuration Manager: It configures the security mechanism and enables end-to-end security between devices.

  - A ZigBee network can adopt one of the three topologies: Star, Tree, and Mesh.

EXERCISES

1. Define big data and big data analytics?

2. What are the big data characteristics?

3. What are the big data processing groups?

4. Draw the Big data flow?

5. What are the big data analytics for security issues and privacy challenges?

6. Define cloud computing?

7. What are the two ends of cloud computing?

8. What are the cloud computing deployment models?

9. What are the three layers of Cloud computing services model and their services?

10. What are the security concerns and challenges of cloud computing?

11. Define Internet of Things and draw its concept topology?

12. What are the building blocks of the IoT?

13. Make map comparison between IoT and M2M?

14. What are the layered models of IoT?

15. What are the applications of IoT for automation, health care, smart manufacturing, and smart cities?

16. What are the new challenges created by IoT?

17. What are the security requirements of IoT?

18. What are the requirements for lightweight cryptography?

19. What are the three primary targets of attack against IoT?

20. What are the five steps for the prevention of attacks on IoT?

21. Define smart grids?

22. What are the requirements of security mechanisms characteristics of the grid environment?

23. What are the challenges of smart grid?

24. What are the smart grid's layers?

25. What are the information security risks and demands of smart grid?

26. Draw the smart grid system classifications?

27. List the possible Cybersecurity attacks in a smart grid?

28. Define SCADA?

29. What are the components of SCADA systems?

30. What are the SCADA system Layers?

31. What are the categories for security threats to modern SCADA systems?

32. Define WSN?

33. What are the WSN Layers?

34. What are the security requirements in WSNs?

35. What are the attacks categories in WSNs?

36. List the attacks and defense in WSNs at different layers?

37. What are the security protocols in WSNs?

38. Define ZIGBEE protocol?

INDEX